Underground Spaces

WIT PRESS

WIT Press publishes leading books in Science and Technology.
Visit our website for the current list of titles.
www.witpress.com

WIT eLibrary

Home of the Transactions of the Wessex Institute.
Papers presented at Underground Spaces 2008 are archived in the WIT eLibrary in volume 102 of WIT Transactions on The Built Environment (ISSN 1743-3509).
The WIT eLibrary provides the international scientific community with immediate and permanent access to individual papers presented at WIT conferences.
Visit the WIT eLibrary at www.witpress.com.

FIRST INTERNATIONAL CONFERENCE ON UNDERGROUND SPACES – DESIGN, ENGINEERING AND ENVIRONMENTAL ASPECTS

Underground Spaces

CONFERENCE CHAIRMEN

C.A. Brebbia
Wessex Institute of Technology, UK

D. Kaliampakos
National Technical University of Athens, Greece

P. Prochazka
Czech Technical University, Prague

INTERNATIONAL SCIENTIFIC ADVISORY COMMITTEE

Organised by
Wessex Institute of Technology, UK

Sponsored by
WIT Transactions on the Built Environment

WIT Transactions

Transactions Editor

Carlos Brebbia
Wessex Institute of Technology
Ashurst Lodge, Ashurst
Southampton SO40 7AA, UK
Email: carlos@wessex.ac.uk

Editorial Board

Underground Spaces

Design, Engineering and Environmental Aspects

Editors

C.A. Brebbia
Wessex Institute of Technology, UK

D. Kaliampakos
National Technical University of Athens, Greece

P. Prochazka
Czech Technical University, Prague

WIT *PRESS* **Southampton, Boston**

Editors:

C.A. Brebbia
Wessex Institute of Technology, UK

D. Kaliampakos
National Technical University of Athens, Greece

P. Prochazka
Czech Technical University, Prague

Published by

WIT Press
Ashurst Lodge, Ashurst, Southampton, SO40 7AA, UK
Tel: 44 (0) 238 029 3223; Fax: 44 (0) 238 029 2853
E-Mail: witpress@witpress.com
http://www.witpress.com

For USA, Canada and Mexico

Computational Mechanics Inc
25 Bridge Street, Billerica, MA 01821, USA
Tel: 978 667 5841; Fax: 978 667 7582
E-Mail: infousa@witpress.com
http://www.witpress.com

British Library Cataloguing-in-Publication Data

A Catalogue record for this book is available
from the British Library

ISBN: 978-1-84564-125-2
ISSN: 1746-4498 (print)
ISSN: 1743-3509 (on-line)

The texts of the papers in this volume were set individually by the authors or under their supervision. Only minor corrections to the text may have been carried out by the publisher.

Printed in Great Britain by Athenaeum Press Ltd.

Preface

This book contains some of the papers presented at the International Conference on Underground Spaces – Design, Engineering and Environmental Aspects, held on the Campus of the Wessex Institute of Technology in the New Forest. The Conference was launched to discuss not only the structural and environmental material characterization aspects but also the trends regarding the development of underground spaces.

Underground spaces are becoming increasingly important for a wide diversity of uses. They range from classical excavations to subway constructions, underground sports halls, power stations, waste repositories, underground cities and many others. The construction techniques are also very varied, from open air excavation to newly developed injection methods.

The use of underground spaces is challenging to a wide spectrum of engineers designers and builders. Structures constructed below the terrain require special attention in their design and safety assessment. This means that the degree of knowledge needs to be significantly different than for surface structures and hence the importance of conferences like Underground Spaces to reach a better understanding of the issues involved. This is particularly the case when preparing to build chemical, nuclear or toxic waste repositories, where serious environmental issues can arise.

The papers presented at the Conference described a variety of problems related to underground spaces. The Editors are grateful to all contributors for their papers and particularly to the members of the International Scientific Advisory Committee who helped to select the material published in this book.

The Editors
The New Forest, 2008

Preface

Contents

Underground space development: setting modern strategies

D. Kaliampakos & A. Benardos
National Technical University of Athens,
School of Mining & Metallurgical Engineering, Greece

Abstract

Underground space development is an irreversible trend especially in urban environments. At this time the underground facilities have proved their usefulness in terms of efficiency and environmental friendliness. Nevertheless, in order to fully exploit the subsurface, new strategies need to be adopted in the whole context of city planning. This includes the introduction of new terms such as the valuation of the underground space, the adoption of integrated planning and zoning policies of the underground uses and the modernisation of the legal framework to incorporate the three-dimensional partition of the property. This paper discusses these issues, the adoption of which can lead to the development of a strategic underground plan, facilitating and further mobilising the hidden potential of underground space utilisation.

Keywords: valuation of underground space, planning and zoning of the subsurface, ownership rights of underground space.

1 Introduction

It has long been recognized that the utilization of the underground space represents a proficient choice to provide solutions to pressing urban problems. Nevertheless, underground projects have been rather focused, until the early 1970s, on the development of transportation infrastructure [1]. Nonetheless, the construction of major transit projects such as metros and road tunnels is just a prelude for the true nature of underground development. The latter encompasses the relocation of several surface land uses or activities, in which installation is difficult, impractical, less profitable, or even environmentally undesirable on the ground level, into subsurface built environments. In the last couple of decades

WIT Transactions on the Built Environment, Vol 102, © 2008 WIT Press
www.witpress.com, ISSN 1743-3509 (on-line)
doi:10.2495/US080011

the engineering community has produced a great deal of exceptional underground projects that served their purpose with an increased efficiency compared to the respective surface solutions and proved the technological capabilities of the construction industry [2, 3]. Even so, these projects have been developed in a passive manner, as they trailed the city's development and aimed at correcting or mitigating the problems caused by the surface expansion. Thus, a great deal of such underground projects were not part of integrated city planning, but they were rather aimed at solving locally existent problems. Hence, until now the true potential of underground structures is yet unexploited and the true challenge in current times is to pass to the mature phase of underground urban utilisation through the incorporation of a strategic multi-disciplinary vision about the use of subsurface space as an integral part of future physical planning and zoning [4].

In the new century, the development of mega-cities, coupled with more pressing needs for a better urban environment, will raise demands for enhanced underground solutions [5–7]. Demands that cannot be met, if either the current planning mode is to be followed without correcting its mistakes or the urban underground resource is consumed without any form of control.

All the above put forth more challenging tasks, like the efficient and sustainable utilization of the subsurface, the adoption of underground solutions - completely replacing several aboveground uses - as the common practice, and the introduction of novel tools capable of measuring the effect that underground space development will have on modern cities. The paper analyzes the new driving forces for strategic subsurface planning and furthermore examines the issues that need to be addressed and adjusted for underground space development so as to meet the requirements of the new era.

2 Urbanisation – the main driving force for underground development

Interest in underground development, especially in urban areas, is constantly increasing internationally. As noted, the main driving force behind the process is the continuously growing urban areas, coupled with the demand for high quality environmental conditions. Unless, one or both these factors cease to exist, the exploitation of the urban subsurface will undoubtedly be in the centre of attention.

One of the most remarkable features of the previous century was the growth of the global population. In the beginning of the 20th century, the world population amounted to 1.6 billion approximately. Today global population is about 6.5 billion and it is estimated that it will reach 9 billion by 2030. Probably more interesting are the development of urban centres and the unparalleled growth of the urban population. In 1800, the percentage of the total population that lived in urban regions was only 3%. It was not until 1820 that London had became the first city that exceeded 1 million residents; in 1900 the number of cities with a population of 1 million residents amounted to 11, whereas the percentage of the total population that lived in urban regions had risen to 14%.

From this point on the development was very fast. In 1950 word population was 2.5 billion and 83 cities had population above 1 million residents, whereas only two cities, London and New York, measured above 10 million residents. The corresponding figures for the urban population are 731 millions in 1950 and 3.1 billion in 2005. It is observed, that while the total population increased by 156% in the last 50 years, the urban population increased by 333% during the same time [8]. Today, 49.2% of world's population live in urban regions. According to a recent UN research [9] it is estimated that the world urban population will reach 4.98 billion by 2030 compared to 2.86 billion in 2000, representing about the 60% of the world population. In the developed countries the percentage of urban population is significantly higher (76%) compared to the developing ones (40%). The latter, however, present higher rates of urbanization. Table 1 presents the forecast for the development of Europe and Northern America's population between 2000 and 2030. It is observed that in 2030 more than 80% of the total population, in both continents, will reside in urban areas. Furthermore, what is worth noticing is that, while the total population of Europe is expected to be reduced by approximately 8%, the urban population is estimated to be increased by roughly 1%.

Table 1: Population growth forecast for Europe and North America [8].

	2000		2030	
	North America	Europe	North America	Europe
Total Population	314.000.000	727.000.000	396.000.000	670.000.000
Urban Population	243.000.000	534.000.000	335.000.000	540.000.000
% urban/total pop.	77,4	73,5	84,6	80,6
% total pop. growth			26,11	-7,84
% urban pop. growth			37,86	1,12

Such has been the growth of urban agglomerations that new terms have been developed to describe the phenomenon as, for example, the term "megacity" that refers to cities with a population of over 5 million residents. Unfortunately all these have come at a cost. It is widely accepted that the lack of free surface space, the sorely high land prices and the deterioration of the environmental conditions are just a few of the repercussions of urbanization [10].

Initially, urban planning opted for the obvious solution, namely spatial expansion, in order to ameliorate these problems. However, as the city's borders were continuously expanding, consuming greedily free space the result proved twofold. On the one hand, the line that distinguishes urban and suburban areas grew thinner and, on the other hand, soon it became apparent that the urban sprawl resulted only in the immigration of the problems to adjacent areas [11]. At this point, sooner or later, depending generally on a country's development a new alternative emerged: the development of underground space. Among the main advantages deriving from the utilization of underground space are the release of space on the surface, the preservation of "sensitive" areas, such as historical city centres, archaeological sites and considerable energy savings. At the same time the installation of hazardous processes (industrial uses, hazardous waste treatment and disposal, etc.) below ground level ensures minimum risk and disturbances (visual impact, noise pollution, odours, etc.) generated by these activities [2].

3 Going underground with a plan

In order to fully exploit the underground the issue of underground land use planning should be brought forward. That means that there should be an overall long-term planning regarding the siting of the underground facilities along with their prioritisation in terms of importance, feasibility and environmental performance. This planning integrated with the possible identification of the development needs can assist in proactively and efficiently construct the right underground structure in the most appropriate place [12]. As a result the optimisation of its positive impacts could be achieved without at the same time jeopardising the misuse or irrational consumption of the underground space.

Of course these policies are not easy to be implemented. Nevertheless a boost in that direction could be made if the following preparatory steps are to be taken:

- The assignment of a value to the underground space.
- The comprehensive investigation and mapping of the underground space, especially in urban areas.
- The adjustment and modernisation of the legislative framework governing the underground space.

3.1 The value of underground space

In the majority of cases underground space is considered to be a public good and a zero value is assigned to it [13]. That means that its "consumption" could take place without paying virtually any cost at all. But let's consider a case where the existence of an underground space in a particular area can have an adverse effect to the construction cost of another underground structure that is proposed to be build adjacent, over or under it. One can argue that the excess cost (e.g. for additional or bypass works, or better support measures) can be considered as the price to pay for the already consumed underground space. This is only a simple example showing that if the value of underground space is ignored, incorrect or misleading assumptions could be made in the planning of the subsurface. Thus it might lead to a non-optimum utilisation, which in turn, could reduce many of the benefits of underground structures.

The urbanisation, as discussed earlier, leads to the development of more underground structures. This increase in the consumption of subsurface resources is gradually transforming the underground medium from a free good into a commodity. Therefore, underground space mandates a value to be assigned at it. Nevertheless, this is a very complicated task, due to the theoretical and practical problems involved. On the surface the methods and techniques for appraising land value and the value of real estate property in general have been long used and coupled with the well-defined proprietary rights. Such methods for appraising underground space value do not currently exist and their absence further aggravates the situation. With the exception of a few cases, where the value of underground space can be inferred, such as in the case of underground car parks, in the majority of underground works the value of the subsurface remains an open question.

As a consequence, more often than not, underground space value appears to be a missing factor in underground development and planning. Ignoring this parameter may seriously delay, in many cases, any underground development. On the other hand, it can lead to an over-consumption of shallow subsurface space, adding more confusion to the often chaotic current conditions, and unjustifiable under-consumption of deeper underground space [13].

Another interesting remark is that many researchers argue that underground space shares many features with non-renewable resources. This is due to the fact that the use of underground space is practically irreversible. Unlike structures aboveground, which can be demolished and rebuilt differently, underground works, in almost all of the cases, cannot be demolished. Underground development changes permanently the existing conditions and there is no realistic way of re-establishing the initial conditions. Therefore, its "consumption" should be done after careful and detailed planning in order for the society to reap the benefits of underground development [14].

As a general rule, the economic feasibility of underground works is judged on the grounds of the comparison between underground and surface construction cost, plus the land cost. However, this comparison reflects only a part of the truth, as the impacts of the planned structures to the environment or the society, expressed in monetary terms are not incorporated in the analysis. Therefore, in order to provide an answer regarding the social benefits of underground solutions, it is necessary to evaluate all the benefits and costs, including the so-called externalities. In other words, underground solutions should be assessed on the grounds of social cost-benefit analysis, using bottom-up approaches and environmental valuation methods. Although there are difficulties in environmental valuation, internationally the use of environmental economics in project appraisal has significantly increased, since it results in better decisions [15].

Towards this direction, both, primary research, based on revealed methods (e.g. Travel Cost and Hedonic Pricing) and preference methods (Contingent Valuation), as well as Benefit Transfer studies have been conducted. Empirical evidences show that the scarcity of free space, the need to protect existing green areas from further degradation and the will to enhance living conditions in modern urban centres tend to increase the cost-effectiveness of underground development and, consequently, their net social benefits.

To illustrate the above with an example let us consider the case of an underground parking facility. This plan allows for a corresponding increase of free space in the surface, which could be developed as an urban recreational green area. According to a research by Damigos and Kaliampakos [16], it was estimated that an urban park of 20,000 m^2 in a densely populated region of Athens affects the dwelling prices at a range between 1 – 4 blocks. Within this zone, a property attracts a premium of 14% up to 31%. More specific, given that the average unit price of an apartment, in the case examined, was about 1,320 €/m^2, the value of the green space capitalized in property prices of the surrounding dwellings ranged between 185 up to 409 €/m^2. It is apparent that in the comparison between an underground facility and the equivalent surface one, apart from the construction, the operational and the land cost, the benefits created by the green areas should also be considered. In this way, if the value of environmental goods and services is taken into account, the advantages of the underground solution would be revealed on strict financial grounds as well.

Nishi et al. [17] presented a similar case in 2000. The authors used a questionnaire in order to establish the residents' Willingness To Pay – WTP to prevent any surface construction that would result in visual degradation of the landscape. The researchers interviewed residents in the cities of Hakodate, Nagoya, Kyoto and Kobe. The results showed that residents were willing to pay $77.5/year/person to preserve the view.

3.2 Mapping and planning of underground space

The vision for underground space development, especially in congested urban areas, mandates for a detailed mapping of the subsurface. Particular attention should be paid to special geological opportunities in terms of rock/soil type and easy access from the surface to the favourable locations. In this issue, there are some notable examples where city or state authorities have collected geological information. More particularly, in Japan, three-dimensional soil-structure data systems have been developed by governmental agencies and are addressed as a part of a GIS [18]. The information, primarily intended for specialists in geology, is also used by engineers in charge of ground surveys. The geological survey of Japan has issued a digital geo-science map, while the Tokyo Metropolitan has carried out geological surveys in conjunction with building construction work and urban base improvement projects. The extensive information from these surveys is incorporated in a relational database and can be retrieved and graphically present borehole, and groundwater data. In Finland, the geo-information is provided by soil and bedrock maps, as well as by the topographical maps provided by the Geological Survey of Finland and local authorities. In Helsinki [19], the Helsinki Geotechnical Database has been established from 1955 and at present contains detailed information on 200,000 site investigation points including sampling and laboratory tests, representing a combined cost of investigation amounting approximately 45 million Euros.

Such geological and geotechnical information contain crucial data nevertheless, it represents only the first step towards the ultimate goal of efficient underground space utilisation. In order to carry forward the development and utilization of underground spaces systematically, a subsurface map that can be used in the development planning and operation stages is necessary. The concept of underground mapping is not to be only limited to geo-information but rather implies that the full resource potential of urban underground must be investigated and recorded. The mapping of underground space should encompass its interaction with the surface buildings and the existing structures along with a land use policy for the underground, as already enforced in typical surface space. Therefore, engineers and urban planner should co-ordinate their actions, aiming at:

- identifying current uses of underground facilities and investigate possible future underground uses and needs in the urban environment
- determining the areas with a high potential for the use of underground space
- recognizing the development scenarios that might encourage or impede the use of underground space.

Stakeholders should decide on the short and long – term urban planning needs and by utilising the data regarding the prevailing geological conditions three-dimensional maps of the subsurface utilisation should be drawn. Consequently, the identification and the promotion of appropriate zoning of several underground use types will be available, along with the time frame for their construction and operation life. Prioritizing the development of conflicting or overlapping underground structures is also of major importance, as well as the recognition of areas where the underground space should be reserved for future needs.

3.3 Ownership of underground space

Legal and administrative restrictions on the development and use of underground space may act as significant barriers to the use of this resource. One of the most significant issues is the proprietary rights of the underground space. Since national territories, local jurisdictions and private ownership are normally defined in terms of boundaries of surface land area, it is necessary for underground space to define how surface ownership extends downwards to the underground and upwards to the sky. Since the roman times, it has been accepted by most western laws that: "*Cujus est solum, ejus est usque ad coelum et ad inferos - To whomsoever the soil belongs, he owns also to the sky and to the depths*" [20].

Since underground space may allow functions to occur within the space independent of the surface land above, questions often arise as to what extent surface land use regulations should apply to the development of underground space. Laws that control land ownership vary among countries, resulting in a state of uncertainty regarding the ownership of subsurface. According to a survey carried out by ITA's WG4 [20] four types of proprietary rights are found to exist:

- unlimited ownership to the centre of the earth
- as far as reasonable interest exists
- only to a limited depth
- the underground is also publicly owned, as private land ownership is almost nonexistent.

The underground space ownership with respect to non-mining activities, especially in urban areas, where several conflicts exist between private and public interests, remains so far unsolved in the majority of the countries. A few of them however have recognised the need to revise their legislation in terms of land ownership and started investigating three-dimensional delimited real estate. For example Oslo (Norway), has already adopted the three-dimensional real estate model [7]. In Japan the land ownership hindered the realisation of major underground public projects and in 2001 a new law was enforced which limited the private land ownership to the depth of 50 m (including deep foundations), while the state owns and manages the subsurface below 50 m, the "deep underground". In the case of the city of Montreal [21] it is widely accepted that the development of the underground city of Montreal would have not been accomplished, unless urban planners had decided to stratify the property rights both vertical and horizontal, etc.

Nowadays, it is more pressing than ever for national governments to update their legal framework related to ownership of subsurface space, so that development of underground infrastructure will be facilitated. The legal responsibilities of owners of underground space and other affected parties should be clearly defined and this can be made possible either by enabling the three-dimensional property or by other amendments (e.g. subsurface volumetric trading).

4 Conclusions

As the trend for underground development firmly established its position in urban planning and the experience gained from underground projects gradually dispersed any doubts concerning the advantages and the superiority of underground structures against contemporary urban problems, building in the subsurface becomes the first choice.

Advances in fields such as rock mechanics, excavation and support of underground structures will undoubtedly enable the construction of more complex and difficult underground projects even in cases when building underground was previously not considered as an option. This constitutes the necessary starting point, still, the strategic vision towards the optimal underground space utilization needs modern strategies to reflect the new prevailing conditions. Issues like the valuation of the subsurface recourse, the mapping and zoning of underground uses and the ownership rights of the underground should be brought forward and resolved in order to have a new, clear framework that will eventually lead to the promotion the subsurface utilisation.

References

[1] Mavrikos, A.A., & Kaliampakos, D.C., Underground development in urban areas: the birth, the evolution and the perspectives of the trend. *Proc. of the 4th Int. Conf. on Urban Regeneration and Sustainability "The Sustainable City"*, Tallinn, Estonia, 17–19 July, 2006.

[2] ITA, *Underground works and the environment*, Report of the Working Group on Underground Works and the Environment, International Tunnelling Association, 1998.

[3] Besner, J., The sustainable usage of the underground space in metropolitan area. *Proc. of the 9th ACUUS Int. Conf. "Underground Space: a Resource for Cities"*, Turin, Italy, 14-16 November 2002.

[4] Ronka, K., Ritola, J., & Rauhala, K., Underground Space in Land-Use Planning. *Tunnelling and Underground Space Technology*, **13(1)**, pp.39–49, 1998.

[5] Damigos, D., Benardos, A., & Kaliampakos, D., The space beneath: Developing the new human-friendly cities. *Proc. 1st Int. Conf. Advances in Mineral Resources Management and Environmental Geotechnology*, 7-9 June, Crete, pp. 641–646. 2004.

[6] Maire, P., Pascal Blunier, P., Parriaux, A., Tacher, L., Underground planning and optimisation of the underground resources' combination looking for sustainable development in urban areas. *Proc. Workshop "Going Underground: Excavating the Subterranean City"*, 21-22 September, Manchester, UK, 2006.

[7] Landahl, G.M., Planning of Underground Space, eds Franzèn T., Bergdahl, S. and Nordmark, A., *Proc. of the Int. Conf. on Underground Construction in Modern Infrastructure*, Stockholm, Sweden, June 7-9, pp. 95–100, 1998.

[8] Geohive, www.geohive.com

[9] UN-HABITAT, 2004/05 Report, State of the world's cities, www.unhabitat.org

[10] Kaliampakos D., & Mavrikos A., Underground Development in Greece: History, Current Situation and Trends. *Proc. 1st Int. Conf. Sustainable Development and Management of the Subsurface*, 5-7 November, Utrecht, 2003.

[11] Mavrikos, A.A., & Kaliampakos, D.C., Appraising the environmental advantages of underground storage facilities in Athens, Greece. *Proc. of the 11th ACUUS Int. Conf., "Underground Space: Expanding the Frontiers"*, 10-13 September, Athens, Greece, pp. 267–272, 2007.

[12] Mavrikos, A.A., & Drakouli E., Incorporating underground space in urban planning. *Proc. of the 11th ACUUS Int. Conf., "Underground Space: Expanding the Frontiers"*, 10-13 September, Athens, Greece, pp. 317–322 2007.

[13] Riera, P., & Pasqual, J., The importance of urban underground land value in project evaluation: a case study of Barcelona's utility tunnel, *Tunnelling and Underground Space Technology*, **7(3)**, pp. 243–250, 1992.

[14] Sterling, R.L., & Godard, J-P. Geoengineering considerations in the optimum use of underground space. ITA-AITES Position Papers, 2001.

[15] Damigos D. & Kaliampakos D., Economic valuation of mined land reclamation: An application of Individual Travel Cost Method in Greece. *Proc. of the International Conference, SGEM 2001: Modern Management of Mine producing, Geology and Environment Protection*, Bulgaria, 2001.

[16] Damigos, D. & Kaliampakos, D., Environmental Economics and the Mining Industry: Monetary Benefits of an Abandoned Quarry Rehabilitation in Greece. *Environmental Geology*, **44(3)**, pp. 356–362, 2003.

[17] Nishi, J., Tanaka, T., Seiki, T., Ito, H., & Okuyama, K., Estimation of the value of internal and external environment in underground space use, *Tunnelling and Underground Technology*, **15(1)**, pp. 79–89, 2000.
[18] Takasaki, H., Chikahisa, H., & Yuasa, Y., Planning and Mapping of Subsurface Space in Japan. *Tunnelling and Underground Space Technology*, **15(3)**, pp. 287–301, 2000.
[19] Vahaaho, I., From geotechnical maps to three-dimensional models. *Tunnelling and Underground Space Technology*, **13(1)**, pp. 51–56, 2000.
[20] Sterling, R., *Legal and administrative issues in underground space use. A Preliminary Survey of Member Nations of the International Tunnelling Association*, International Tunnelling Association, 1990.
[21] Escobar, M., The Next Urban Frontier - The Inner City and the Role of the Evolution of Real-Property Law in the 21st Century - A Montreal Perspective, *Proc. of the 9th International Conference, Urban Underground Space: a Resource for Cities*, 14-16 November, Turin, Italy, 2002.

Blast impact on structures of underground parking

P. P. Procházka[1], A. N. Kravtsov[2] & S. Peskova[1]
[1]*Czech Association of Civil Engineers, Prague, Czech Republic*
[2]*Czech Technical University in Prague, Civil Engineering, Structural Mechanics, Czech Republic*

Abstract

In big cities underground spaces are built up for subways, underground parking and tunnels, etc. These rooms are threatened by terrorist attacks and not only human lives can be lost but also extensive material damage can be expected. This is why it is of great importance to predict dynamic impacts of explosives, which can then be transformed to static statistically evaluated loading. In this paper the impact of explosion and air strike wave is formulated and solved. Gas dynamics and dynamic response of soil with process of dissipation of air-strike energy are considered. This means that some part of this energy is transferred into structures and soil mass. It appears that for contact explosions on the soil surface this part can be up to 30% of the total explosion energy (in soft soils). The variables to be calculated are mass density of gas, the velocity of movements and the internal energy. The latter covers the influence of the gas pressure, being given for the adiabatic state. The air is linearly related to the internal energy of a unit mass of the gas, and the density, while in the neighborhood of the source of explosion the pressure changes nonlinearly with respect to the gas density. Time dependent finite element solution is compared with results published in Lucy, L.B. (1977). A numerical approach to testing of the fission hypothesis. *Astron. J.* **82**, 1013.
Keywords: underground parking, striking wave due to explosion, time development of gas pressure, impact load.

1 Introduction

Numerical simulation of gas explosion is very particular as the system of equations describing the process are nonlinear and of the first order. Moreover,

WIT Transactions on the Built Environment, Vol 102, © 2008 WIT Press
www.witpress.com, ISSN 1743-3509 (on-line)
doi:10.2495/US080021

movement of interfacial boundary between subdomain simulating neighborhood of explosion and the virgin air (gas) moves according to the current situation. Consequently, large movements are expected. In order to describe such displacements of the gas couple of numerical approaches exists. One of the most suiting appears smooth hydrodynamics particle method (SHP), which was historically developed for astrophysical purposes, [1, 2]. The inherent benefit of the SHP formulation consists in transformation of partial differential equations to a system of linear algebraic using regularization. This transformation is, among others, suitable for parallel computations. Recently, SHP has grown into a successful and respected numerical tool. In particular, this method does not differ between 3D, 2D and 1D problems, as the problems defined in higher order spaces can be simulated as easy as that in 1D. An excellent review of the advantages and recent progress in SHP can be found in [3, 4]. Some problems occur when geometrical boundary conditions should be involved. Authors of [5] proposed the ghost particle method, in which some particles are located outside the domain. Heat conduction problem is solved in [6], where Taylor series expansion approximates the regularization kernels.

This paper partly starts with ideas of Veselovsky and Kurepin [7], where the problem of explosion in underground parking is solved. More detailed analysis is submitted in [8], where two-dimensional problems of gas dynamics are comprehensively discussed. The general approach for hydrodynamic processes involving strike waves and high temperature can be found in [9]. This information is collected into a formulation of loading acting against fixed walls of underground parking. The solution of the problem is done in terms of SHP.

2 Methodology of load calculation

The general problem of definition of loading on structures due to explosion and air strike waves is a complicated topic of solid mechanics. It covers a combined solution of gas-dynamics, dynamics of soil and building structures involving processes of dissipation of air-strike energy. That means that some part of this energy is transferred to structures and soil massive. For contact explosions on soil surfaces this part can be up to 30% of explosion energy (on soft soils). But in this case with lifted charge (center of explosion) the problem can be formulated in a simpler way.

Definition of the load will be split into two steps, which are based on gas dynamical calculation. We do not examine processes of transfer of air strike wave energy to soil and structures (parking columns, ceiling, ground, and side walls) and do not contemplate their movements into calculation, i.e. the boundaries of the air space are stiff.

In the first step we calculate the beginning stage of the process of impacts of explosion and spread out of the air strike wave until the moment of contact with the structures.

Calculation of strike wave parameters at the beginning stage is base on numerical calculation of one dimension equations of gas dynamic. Distribution of density $\rho(\boldsymbol{r})$, velocity $\boldsymbol{u}(\boldsymbol{r})$ and internal energy $\varepsilon(\boldsymbol{r})$ ($\boldsymbol{r}$ is the radius or distance from the origin, which is centered at the point of explosion) at the moment of beginning of the interaction of the air strike wave with nearest structures used for calculation of spread out and various interactions among air strike waves are incorporated in the interfacial conditions with the second stage.

In the second step processes of interaction of the air strike waves with structures are studies. Strike wave parameters appearing in the second step are computed from equations of three-dimensional gas dynamic.

3 Equations of motion, calculated parameters, quantities and their dimensions

Mathematical modeling of the air movements is base on the solution of equations of gas dynamics, which for three-dimensional problem in Cartesian system of coordinates are listed as:

$$\frac{\partial \rho}{\partial t}+\frac{\partial(\rho u_x)}{\partial x}+\frac{\partial(\rho u_y)}{\partial y}+\frac{\partial(\rho u_z)}{\partial z}=0 \tag{1}$$

$$\frac{\partial(\rho u_x)}{\partial t}+\frac{\partial(p+\rho u_x^2)}{\partial x}+\frac{\partial(\rho u_x u_y)}{\partial y}+\frac{\partial(\rho u_x u_z)}{\partial z}=0 \tag{2}$$

$$\frac{\partial(\rho u_y)}{\partial t}+\frac{\partial(\rho u_x u_y)}{\partial x}+\frac{\partial(p+\rho u_y^2)}{\partial y}+\frac{\partial(\rho u_y u_z)}{\partial z}=0 \tag{3}$$

$$\frac{\partial(\rho u_z)}{\partial t}+\frac{\partial(\rho u_x u_z)}{\partial x}+\frac{\partial(\rho u_y u_z)}{\partial y}+\frac{\partial(p+\rho u_z^2)}{\partial z}=0 \tag{4}$$

$$\frac{\partial e}{\partial t}+\frac{\partial[(e+p)u_x]}{\partial x}+\frac{\partial[(e+p)u_y]}{\partial y}+\frac{\partial[(e+p)u_z]}{\partial z}=0 \tag{5}$$

where:

x, y, z Cartesian coordinates [m]

u_x, u_y, u_z components of the vectors of velocity U, [m/msec]

$|U|^2 = u_x^2 + u_y^2 + u_z^2$ norm of the vector of velocity

$\rho = \rho(x,y,z,t)$ density of gas [kg/ m^3]

$p = p(x,y,z,t)$ pressure of gas [MPa]

$e = \rho[\varepsilon - (u_x^2 + u_y^2 + u_z^2)/2]$ full energy of a unit of mass of the gas, [MPa]

$\varepsilon = \varepsilon(x,y,z,t)$ potential energy [m^2/ms^2]

$\frac{1}{2}(u_x^2 + u_y^2 + u_z^2)$ kinetic energy [m^2/ms^2]

Equation describing explosion of TNT charge can be recorded as

$$p = (\gamma - 1)\rho\varepsilon, \tag{6}$$

where γ is the exponent of adiabatic process. For the air γ = 1.4, in case of explosion the exponent of adiabatic process becomes density dependent, i.e. $\gamma = \gamma(\rho)$. Exponent of adiabatic process $\gamma(\rho)$ can be calculated in the following way

- $\gamma = 3$ if $\rho > 440\ \mathrm{kg/m^3}$;
- $\gamma = 1.3$ if $\rho < 50\ \mathrm{kg/m^3}$;
- $\gamma = \gamma(\rho)$ if $50 \le \rho \le 440\ \mathrm{kg/m^3}$ - linear interpolation

can be applied (monotonic and smooth dependence on density ρ is assumed).

Using matrix notation,

$$\boldsymbol{\sigma} = \begin{Bmatrix} \rho \\ \rho u_x \\ \rho u_y \\ \rho u_z \\ e \end{Bmatrix}, \quad \boldsymbol{a} = \begin{bmatrix} \rho u_x \\ p + \rho u_x^2 \\ \rho u_x u_y \\ \rho u_x u_z \\ (e+p)u_x \end{bmatrix}, \quad \boldsymbol{b} = \begin{bmatrix} \rho u_y \\ \rho u_x u_y \\ p + \rho u_y^2 \\ \rho u_y u_z \\ (e+p)u_y \end{bmatrix}, \quad \boldsymbol{c} = \begin{bmatrix} \rho u_z \\ \rho u_x u_z \\ \rho u_y u_z \\ p + \rho u_z^2 \\ (e+p)u_z \end{bmatrix}.$$

the above equations can be recorded in a simpler form as $\frac{\partial \sigma}{\partial t} + \frac{\partial a}{\partial x} + \frac{\partial b}{\partial y} + \frac{\partial c}{\partial z} = 0.$

If the formulation possesses certain kind of symmetry three dimensional equations can be transformed into one dimensional equations of the form:

$$\frac{\partial \rho}{\partial t} + \frac{\partial (\rho u)}{\partial r} + \frac{(\nu - 1)\rho u}{r} = 0 \tag{7}$$

$$\frac{\partial (\rho u)}{\partial t} + \frac{\partial (p + \rho u^2)}{\partial r} + \frac{(\nu - 1)\rho u^2}{r} = 0 \tag{8}$$

$$\frac{\partial e}{\partial t} + \frac{\partial [(e+p)u]}{\partial r} + \frac{(\nu - 1)[(e+p)u]}{r} = 0 \tag{9}$$

where r is the space coordinate, ν is a sign of symmetry (ν = 1 - plane, 2 - cylindrical, 3 - spherical symmetry). In case of cylindrical symmetry axial coordinate z is not considered in the above formulas. The nonlinear equations are solved by the method of Godunov *et al.* [8] using special linearization of the above equations.

4 Regularization of functions and their derivatives

The concept behind SHP is based on an interpolation scheme. From mathematical calculus it is well known, [7], that for each generalized function f defined on a domain $V \subset R^n$ with boundary S there exists a positive ε and a finite cover $\{\Omega\}_{i=1}^N \subset V, i = 1,...,N$ (for each point $x \in V$ there is an index $i \in 1,...,N$ so that $x \in \Omega_i$) with measure of $\Omega_i < \varepsilon$ so that on Ω_i there exists function $\omega_\varepsilon^i \in C^\infty(\overline{\Omega})$, supp $\omega_\varepsilon^i \in \Omega_i$ (sometimes called cap function) which regularize the function f in such a way that f can be expressed as

$$f(\boldsymbol{x}) = \sum_{i=1}^{N} \int_{\Omega_i} f(\xi)\omega_\varepsilon^i(\boldsymbol{x} - \xi)\,\mathrm{d}\xi = f * \omega_\varepsilon^i, i = 1,...,N, \tag{10}$$

and the left hand side of the latter relation is called the regularization, $f * \omega_\varepsilon^i$ is the convolution. Recall some basic properties of the regularization: the volume of each cap function is unity, is equal to one. If the function f is uniform (equal to one) and $\varepsilon \rightarrow$ infinity the regularization turns to be density of the function f, for example density of probability. If $\varepsilon \rightarrow 0$ the kernel ω_ε^i turns to be the Dirac function. For each positive ε the regularization (kernel, cap function) ω_ε^i can be created infinitely differentiable (for definition of types of cap functions, see [7], for example).

Since different cap functions should be created for different Ω_i, the above definition becomes inconvenient. In order to improve this put $\omega_\varepsilon^i \equiv \omega_\varepsilon$ and the shape of Ω_i remains same for all i, the area of a circle in 2D or the volume of a sphere, for example. Now inside of the domain V select a set of points $\boldsymbol{x}_i, i = 1,...,N$, $\boldsymbol{x}_i$ is centered at Ω_i and a new function F is defined as

$$F(\boldsymbol{x}_i) = \sum_{i=1}^{N} \int_{\Omega_i} f(\xi)\omega_\varepsilon(\boldsymbol{x}_i - \xi)\,\mathrm{d}\xi, i = 1,...,N \tag{11}$$

which is formally similar to relation (1), so that it fulfils basic properties above mentioned. Since the former assumptions take place the function F cannot be expected to be equal to f any longer, but a special case: $\varepsilon \rightarrow 0$ in the sense of definition of the Dirac function.

In our case 2D problem is considered and degrees of freedom are concentrated at nodes $\boldsymbol{x}_i \in \Omega_i, i = 1,...,N$, Ω_i are considered as areas of the circles in which x_i is centered. In the approximation, the smoothed (regularized) function F for any physical quantity f is identified with the original function, i.e. $F \equiv f$. Moreover, the kernel ω_ε is simplifies for real calculations and the simplification is denoted as W_ε. Introducing this to (2) and setting $f_i = f(\boldsymbol{x}_i)$ gives:

$$f_i = f(\boldsymbol{x}_i) = \int_{\Omega_i} f(\xi) W_\varepsilon(\boldsymbol{x}_i - \xi)\mathrm{d}\xi \tag{12}$$

Equation (3) is the kernel representation to average functional distribution. In our next considerations additional properties of W_ε will be required:

positivity: $W_\varepsilon(\boldsymbol{x}_i - \xi) \geq 0, \quad \xi \in \Omega_i$

normalization $\int_{\Omega_i} W_\varepsilon(\boldsymbol{x}_i - \xi)\mathrm{d}\xi = 1, \quad \forall \varepsilon > 0$

surface smoothness on $\partial\Omega_i$: $W_\varepsilon(\boldsymbol{x}_i - \xi) = \nabla W_\varepsilon(\boldsymbol{x}_i - \xi) = \nabla\nabla W_\varepsilon(\boldsymbol{x}_i - \xi) = 0, \quad \xi \in \partial\Omega_i$

The last property follows from the fact that the order of differential equations, which are to be studied, is two, and so is the required regularity (continuity).

Using integration by parts, from the boundary conditions on $\partial\Omega_i$ it immediately follows that

$$\int_{\Omega}\nabla W_{\varepsilon}(\boldsymbol{x}_i-\xi)\mathrm{d}\xi=\int_{\Omega}\nabla\nabla W_{\varepsilon}(\boldsymbol{x}_i-\xi)\mathrm{d}\xi=0 \tag{13}$$

For the sake of simplicity the approximation of the kernel W_ε is represented by

$$W_{\varepsilon}(\boldsymbol{x}_i-\boldsymbol{\xi})=C(1-10r^3+15r^4-6r^5),\quad r=|\boldsymbol{\xi}-\boldsymbol{x}_i|/h \quad \text{for 2D} \tag{14}$$

where $C=\dfrac{1}{h}$ for 1D problem, $C=\dfrac{7}{3\pi h^2}$ for 2D problem, $C=\dfrac{42}{5\pi h^3}$ for 3D problem, and $|\boldsymbol{\xi}-\boldsymbol{x}_i|\le h$ is the distance between the pertinent points.

If we consider volume (area, interval) of an element $\Omega_i=\dfrac{m_i}{\rho_i}$, where m_i is the mass of the element and ρ_i is the density, using rectangular rule of evaluation of integrals yields:

$$f_i=f(x_i)=\sum_{r_{ij}\le h}\frac{m_j f_j}{\rho_j}W_{\varepsilon}(r_{ij}),r_{ij}=|\boldsymbol{x}_j-\boldsymbol{x}_i|,$$
$$\nabla f_i=\rho_i\sum_{r_{ij}\le h}m_j\left(\frac{f_i}{\rho_i^2}+\frac{f_j}{\rho_j^2}\right)\nabla W_{\varepsilon}(r_{ij}) \tag{15}$$

5 Calculation schemes

Typical explosive scheme of three dimensional problem is considered in this section. It starts with the position of the charge near the neighboring side walls. This case is depicted in Figs. 1, where in the left picture view of the situation and in the right picture the plot of the situation is seen. The ball centered at the charge position with radius R_c describes the domains of charge in the picture.

The charge position as well as the side walls, ground and ceiling are imbedded in Cartesian coordinates $0xyz$, where the plane $0xy$ is the ground and z is upwards oriented. In both cases the ground is characterized by the plane $z=0$ and the ceiling is the plane $z=3$. The length dimensions are measured in meters. Center of the charge possesses the coordinates ($0.5+R_c$, $0.5+R_c$, $0.5+R_c$) in the first case and in the second case vertical coordinate is $0.5+R_c$, while the other coordinates are zero. Values of the radii of charge and its mass q are introduced in Table 1 for the density of TNT $\rho_{\mathrm{TNT}}=1620\ \ \mathrm{kg/m^3}$.

In the first stage of definition of loads numerical solution of the problem with the air explosion charges of mass $q=50$ kg and $q=100$ kg TNT till the front $r_0=0.5+R_c$.

Characterization of the charge is done by definition of mass and energy inside of the domain of charge. Density of TNT is considered as 1620 kg/m^3.

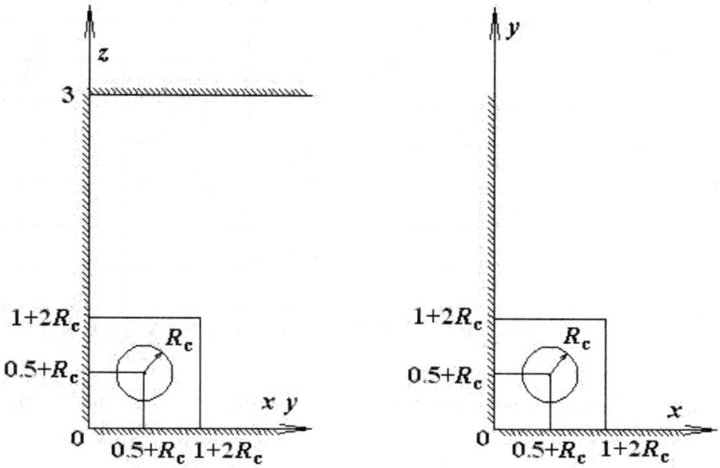

Figure 1: Scheme for the charge position near the sidewalls.

Table 1: Dependency of the radius R_c on the power q.

q, kg	R_c, m
50	0.245
100	0.195

Table 2: Remaining pressure Δp_f.

q, kg	R_f, m	Δp_f, kgs/cm^2
50	0.695	131
100	0.745	156.7

6 Results

Results of calculation – beginning distribution of density $\rho(r)$, speed $u(r)$ and pressure $p(r)$ behind the air strike wave are shown in Figs. 2–7 for the case depicted in Fig. 1. In the graphs of density distribution $\rho(r)$ drop of this function behind the air strike wave is seen. Density disconnection is equal to contact drop that divides influences of charge and the air compressed by the air strike wave. The boundary conditions on the interface of the air and the structures of the parking are prescribed in such a way that fully reflexive surfaces of the structures are considered.

In Table 2 the remaining pressure Δp_f on the front of air strike wave at the beginning of interaction of the air strike wave and the structure for the first case of geometry, see Fig. 1.

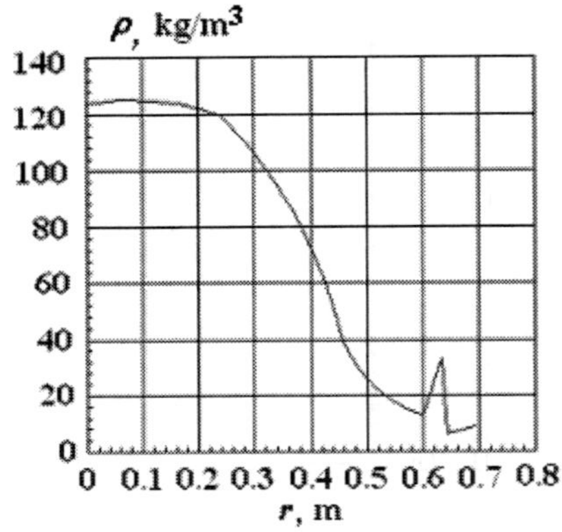

Figure 2: Initial density for q = 50.

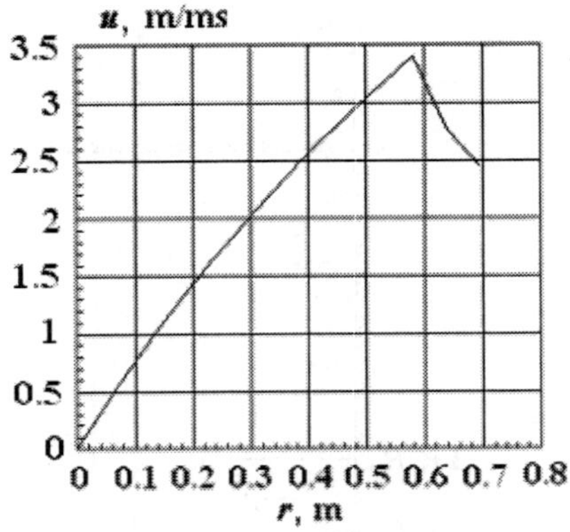

Figure 3: Initial velocity for q = 50.

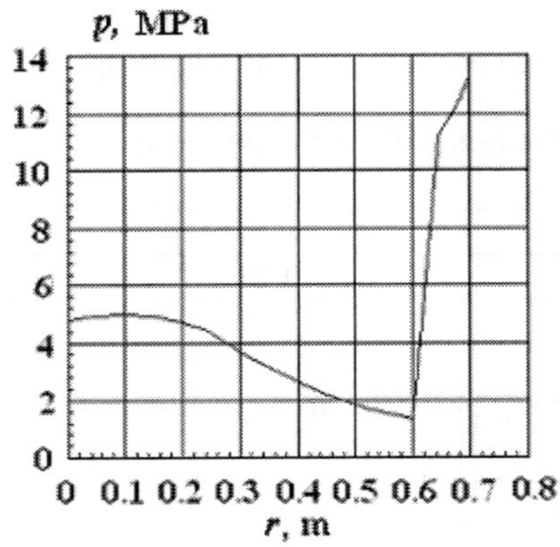

Figure 4: Initial pressure for q = 50.

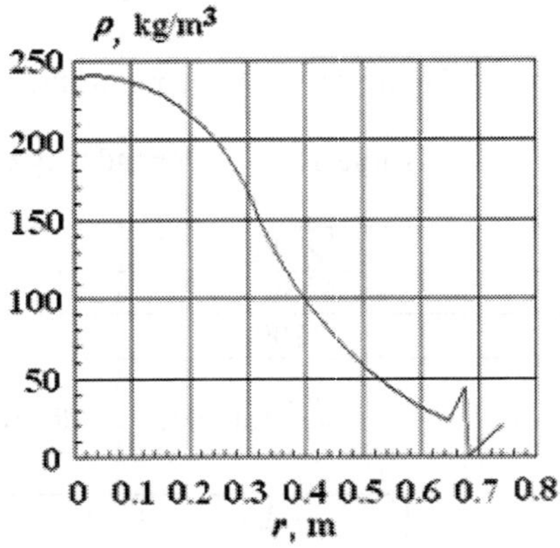

Figure 5: Initial density for q = 100.

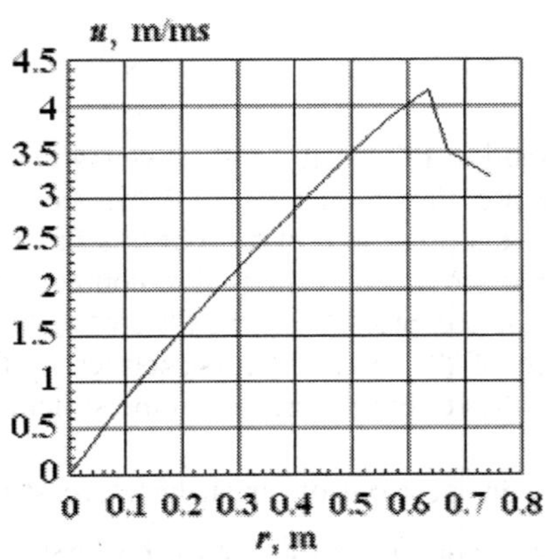

Figure 6: Initial velocity for q = 100.

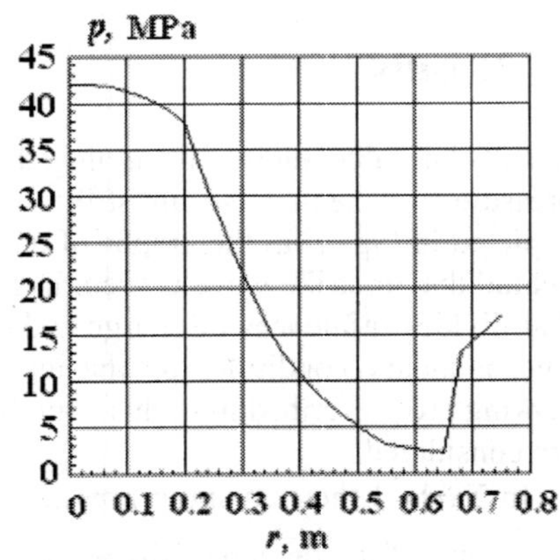

Figure 7: Initial pressure for q = 100.

7 Conclusions

In this paper movements of gas due to explosion in an underground parking are studied. The main purpose of this paper is to discover the loading developed against the side walls, ground and ceiling of the parking room. The SHP method is used as a numerical tool, solving the set of equations describing the movement process of gas (air). Density disconnection appears at the walls due to difference in influences of the charge and the air, which is compressed by the air strike wave. In the neighborhood of the charge supersonic velocity is considered, which induces subsonic velocity in the air.

Acknowledgements

This paper was prepared under financial support of GAČR, project No. 103/08/0922 and MSM, project number 6840770001.

References

[1] Lucy, L.B. (1977). A numerical approach to testing of the fission hypothesis. *Astron. J.* **82**, 1013.

[2] Gingold, R.A. and Monoghan, J.J. (1977). Smooth particle hydrodynamics: theory and application to non-spherical stars. *Monthly Nat. R. Astron. Soc.* **181**, 375.

[3] Randle, P.W. and Libersky, L.D. (1996). Smooth particle hydrodynamics: some recent improvements and application. *Appl. Mech. Engng.* **139**, 175.

[4] Li, S. and Liu, W.K. (2002). Meshfree and particle method and their applications. *Appl. Mech. Rev.* **55**, 1.

[5] Takeda, H., Miyama, S. and Sekiya, M. (1994). Numerical simulation of viscous flow by smoothed particle hydrodynamics. *Prog. Theor. Phys.* **92**, 939.

[6] Chen, J.K., Beraun, J.E. and Carney, T.C. (1999). A corrective smooth particle method for boundary value problems in heat conduction. *Int. J. Numer. Methods Engrg.* **46**, 231.

[7] Veselovsky, A.N., Kurepin, N.S.: Proceedings of 26th Central Scientific Institute of the Russian Federation. Moscow 2006, report II/2, in Russian

[8] Godunov, S.K., Zabrodin, A.V. et.al: Numerical solutions of the poly-dimensional problem of gas dynamic. Moscow, Nauka, 1976, in Russian.

[9] Zeldovich, J.B., Raize, J.P.: Physics of strike waves and high temperature hydrodynamic processes. Moscow, Nauka, 1966, in Russian.

Artificial intelligence in underground development: a study of TBM performance

A. Benardos
National Technical University of Athens,
School of Mining & Metallurgical Engineering, Greece

Abstract

Modelling tunnel boring machine (TBM) performance is an important aspect in tunnel operations. The use of artificial intelligence techniques such as artificial neural networks has been recently introduced to this subject and the results from such applications prove their potential in making accurate prognosis. This paper presents a review of feed-forward artificial neural network (ANN) development and furthermore it illustrates their application by the use of two cases studies from Italian and Greek underground projects, where the TBM performance is modelled. The results obtained show that the developed ANNs can efficiently generalise the TBM behaviour in their respective geotechnical environment, having a reliable, effective and consistent performance.
Keywords: TBM performance modelling, artificial neural networks.

1 Introduction

Assessing the performance of tunnelling operations is one of key data for the overall success of the project as this issue is directly interconnected with the financial performance of the construction works [1]. Even more important is to estimate the tunnelling rate of tunnel boring machines (TBMs), as the flexibility limitations these particular machines have can lead to considerable downtime. These problems are more intense in tunnelling projects constructed in complex geological formations [2] and especially in urban areas where the low construction depth and the external loading from the buildings increase risk conditions [3].

In order to assess the performance of TBMs many researchers have proposed various methodologies [4–12] in an effort to express the penetration rate using as

WIT Transactions on the Built Environment, Vol 102, © 2008 WIT Press
www.witpress.com, ISSN 1743-3509 (on-line)
doi:10.2495/US080031

inputs data relating to the rock mass properties and/or machine characteristics. Beyond mathematical formulae and analytical solutions, artificial intelligence systems and more particularly artificial neural networks (ANNs) have not been introduced in this issue until recently. Nevertheless, many researchers [13–17] have demonstrated very promising results. This is because ANNs can further enhance the effectiveness of the analysis, especially in rock engineering applications such as the one described, where the interrelated parameters are numerous; their interaction is not clearly identified, or is very complicated to be explicitly expressed.

This paper gives a brief review regarding ANN development and furthermore it deals with the modelling of the TBM performance emphasising the identification of the performance oscillations throughout the tunnelling period. This is made possible by the development of ANNs capable of learning from the tunnelling experience and generalising solutions for new sets of input data. Hence, the main aim is to produce a tailor-made model, utilised during the construction period, capable of providing estimates of the expected tunnelling advance rate. To illustrate the efficiency and accuracy of the ANN generalisation two case studies are presented in Italian and Greek underground projects, where the TBM penetration rate is modelled with respect to the geological and geotechnical conditions, as well as the machine characteristics by the use of trained neural networks.

2 Artificial neural networks

2.1 Definition

The development of ANNs started as an attempt to understand the operation of the human brain and mimic its assessment capabilities. In other words, to be able to decide and act under uncertainty or even deal with situations having limited previous experience. ANNs are mathematical models consisting of interconnected processing nodes (neurons) under a pre-specified topology (layers).

Neural networks have a strong similarity to the biological brain and therefore a great deal of their terminology is borrowed from neuroscience. Their basic characteristic is the ability to perform massively parallel computing of the input stimulus (data), contrary to the custom mathematical models that are based rather on a serial process of mathematical and logical functions [18]. Another advantage of the ANNs is their flexibility in data processing, as no deterministic mathematical relationship of the examined components is required. Instead, once the data is introduced, in a cause–effect mode, the network identifies the existing relationships, learns and mimics their behaviour by adjusting the strength of the links between the neurons (connection weights). Thus, they cannot be programmed but they are rather taught through case experience. As a result, soon after the ANN's training, given an existing dataset, estimates can be drawn for another specific data input. Thus, the trained network can generalise and give estimates for uncertain conditions or even incomplete data [19]. The main disadvantage of ANNs is that an explicit determination of the parameter's weighting is not an easy task or it may not even be possible in large and complex network architectures. The ANN operation is based on the following:

- Data processing occurs in a number of simple processing units (neurons), which have signal inputs and outputs.
- The neurons' bonding is made through connection links, each one of them having a corresponding weight that multiplies the signal.
- Each neuron applies an activation function to the signal input to control the signal output.

2.2 ANN architecture and training process

In general, a typical ANN topology is consisted by a set of layers; the input layer, one or more hidden layers and the output layer, each one of them containing a certain number of neurons. Accordingly, each neuron is linked to neighbours with varying coefficients of connectivity that represent the weighting of these connections. Each neuron of the hidden layer(s) is interconnected to all others found in the input and output layers.

The type of ANN used in this paper are the feed-forward neural networks, which are the most widely used. They are commonly applied to problems where a set of input vectors should be corresponded to another specified set of output vectors. The training procedure consists of a sequential data feed into the network, followed by the comparative evaluation of the corresponding output provided by the ANN and the actual result. The network adjusts the weighting of the connection links in the neurons of the hidden layers in a continuous effort to produce the results that would best correspond to the training dataset. A complete pass of all the input data through the network consists a training epoch and usually a great number of epochs is required for the residual error to converge below a pre-specified threshold. A schematic illustration of a feed-forward ANN training is given in fig. 1.

Feed-forward ANNs are usually trained with the backpropagation algorithm, also known as the generalized delta rule. In order to train a feed-forward ANN, corresponding sets of input (training input vectors) and output (target output

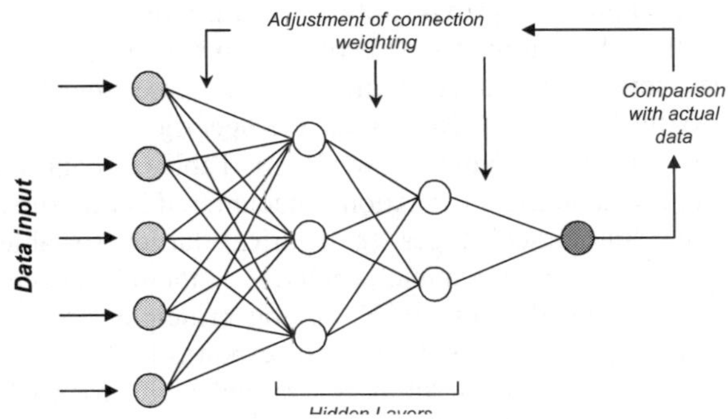

Figure 1: Training process of a feed-forward ANN with two hidden layers.

vectors) data must be presented to it. Each target output vector is the ANN's desired response to the appropriate training input vector. The training algorithm is used to modify the connection weights so as to minimize the error between the ANN's output and its desired response for all training input vectors. Generally, the training error is a function of the difference between the ANN's predicted values and desired responses, with the connection weights as the independent variables. Common formulations used for the training error include the sum squared error (SSE) and the mean squared error (MSE).

3 ANN in TBM performance prediction – cases studies

The whole idea follows the ANN philosophy, that is, to analyse the experience gained from the tunnel boring process and to correspond it to a set of selected data. This cause-effect request is used in the ANN so as to identify the interactions between the data and to come up with the appropriate weighting of the parameters involved, which will finally determine the generalisation accuracy.

In order to illuminate the ability of ANNs to generalise solutions depicting the TBM performance, two case studies are presented. The first one deals with two Italian tunnels (Maen and Pieve), where the penetration rate is modelled based on input data relating to ground properties and machine characteristics. The second is related to an interstation tunnel, from the Athens metro project, where an ANN is trained based on a series of geotechnical data in order to be able to reveal possible risk prone areas where TBM operation is negatively affected by ground conditions. All cases are modelled individually, as the geotechnical environment, as well as, the particular characteristics of each TBM used are different. Thus, the development of separate ANNs enhances the precision and the efficiency of the generalisations that could be further used in order to have consistent prognosis for the corresponding geotechnical setting.

3.1 Case study 1 – Italian tunnels

Data for TBM performance analysis have been obtained from two tunnels (Maen, Pieve) excavated in metamorphic rocks located in the Italian Alps. The combined tunnel length is approximately 11.5 km, while data records exist for the 8.5 km. In the Maen tunnel the recordings were made at a 5 m interval, whereas at the Pieve tunnel data relating to the geotechnical conditions was gathered on a daily basis [10]. Details on the specific tunnel projects and the TBMs used are given in Table 1, while in Table 2 data relating to the characteristics of the geological formations encountered is presented.

Regarding the lithological types (categorical target variables) that were introduced in the ANN, for each tunnel, each one of them has been corresponded to an input neuron using the "one-of-c" coding principle. That means that the coding of c binary target variables (0 or 1) corresponds to the c categories. These new variables are also known as "dummy" variables and for each one the zero value is assigned to it, except for the one corresponding to the correct category, which is given the value one. Thus, only the neuron that corresponds to the

Table 1: Construction data for the tunnels under investigation [10].

	Maen	Pieve
Surveyed section length (m)	1750	6400
Total excavation time (days)	413	809
Excavated diameter (m)	4.20	4.05
Tunnel slope (°)	24–35	≈0
TBM model	Wirth 340/420 E	Robbins 1111-234/3
TBM type	Open	Double shield
Number of cutters	36	27
Cutter diameter (in)	17''	17''
Maximum trust (kN)	7920	4602
Boring stroke (m)	1.5	0.63
Cutterhead rotation rate (rpm)	5.5–11	11.3

Table 2: Main characteristics of the geological formations [10].

Tunnel	Rock type	UCS (MPa)	Tensile strength (MPa)	Mean Mohs' hardness	Knoop hardness (GPa)	Cutter Life Index	Young's modulus (GPa)
Maen	Serpentinite	124	—	3.6	—	30–70	—
	Metabasite	180	15	6.2	6.2	10–20	65
	Chlorite schist	17	—	2.8	—	60–90	—
	Metagabbro	138	10–12	6	5.1	15–25	39
	Calc schist	75	—	3.6	—	30–70	—
Pieve	Micaschist	124–215	5–9	4.1	5.2–8.5	15–70	28
	Metadiorite	171–221	8–13	5.1	6.2–7.0	15–40	46–100
	Meta quartzdiorite	160–210	—	6.4	—	15	—
	Metagranite	146–296	0.7–7	6.6	7–10	10	24–38

Table 3: "One-of-c" coding used for the lithologies in the Maen tunnel case.

Maen	SP	Serpentinite	1	0	0	0	0	0
	CHLSC	Chlorite schist	0	1	0	0	0	0
	TALC	Talc schist	0	0	1	0	0	0
	CLS	Calc schist	0	0	0	1	0	0
	MBAS	Metabasite	0	0	0	0	1	0
	MG	Metagabbro	0	0	0	0	0	1

actual encountered lithological type, for the given data array, is activated each time. The "One-of-c" coding used for the lithologies in the Maen tunnel case is given in Table 3.According to the above, for the Maen tunnel there were 8 input neurons (6 for the lithological types), while for the Pieve tunnel the number of the input neurons were 7 (5 for the lithological types).

The datasets from these two tunnels have been discerned into 3 subsets using a uniform sampling process; the training, the testing and the validation ones

From the 330 datasets for the Maen case and the 301 datasets available in the Pieve tunnel, about 60% was used for training, whereas the testing and validation subsets each amounted approximately 20% of the data. The training dataset is introduced to the ANN so as to properly adjust the weighting connections of the neurons against target output (see fig. 1), while the validation subset is used as a barrier to avoid data overfitting, as it stops the training when designated error levels are reached. Finally, the testing subset is used as the measure of evaluating the trained model's efficiency. The input data of this subset are unknown to the model as they are used only after the completion of the training process. The comparison of the model's estimates with the actual output data, documents ANN's ability to generalize (predict). The ANN's performance is assessed in terms of the relative error level (Δ) achieved, between the actual (PR_{actual}) and the predicted penetration rates ($PR_{predicted}$), following the expression:

$$\Delta = \frac{PR_{actual} - PR_{predicted}}{PR_{actual}} \quad (1)$$

This criterion can provide a clear aspect regarding the ANN behaviour and moreover makes possible the comparison between the ANN results and other methods or theoretical models focusing on advance rate prediction.

In both cases, the optimal results were obtained by utilizing two hidden layers, with an increased number of neurons in the first of them. In Table 4, the optimum ANN architectures for the two tunnel cases are given, along with the mean squared errors (MSE) of the training process and the relative error levels (Δ) for the generalisation outputs. The most efficient behaviour is achieved in the ANN developed for the Maen tunnel, having an 8x9x5x1 architecture. This particular structure type means that the ANN has a total of 4 layers, with 8 neurons in the input level, same as the number of the parameters, two hidden layers with 9 and 5 neurons respectively, followed by 1 neuron in the output layer that eventually generates the value of the penetration rate.

Table 4: ANN training and testing error for each examined tunnel.

	Maen	Pieve
Optimum ANN architecture	8×9×5×1	7×6×5×1
Training MSE	0,119	0,086
Relative error of ANN generalisation (%)	17,9%	21,5%

Beyond the presentation of the mean values for the relative error levels it is of equal importance to evaluate the overall behaviour of the trained networks. This will assure that the ability of the ANNs to provide reliable prognosis is spread throughout the dataset and not only focused in particular sections. This check can be made with the use of fig. 2, where the actual penetration rate for the Maen tunnel are presented in conjunction with the ANNs' output for all data incorporated in the testing subset, along with an additional bar-graph presenting the attained relative error. Furthermore, in fig. 3 the scatter plot between actual and modelled penetration rate is given.

All the above concur that the ANNs' generalisations present a satisfactory approximation level, consistent throughout the dataset examined, and

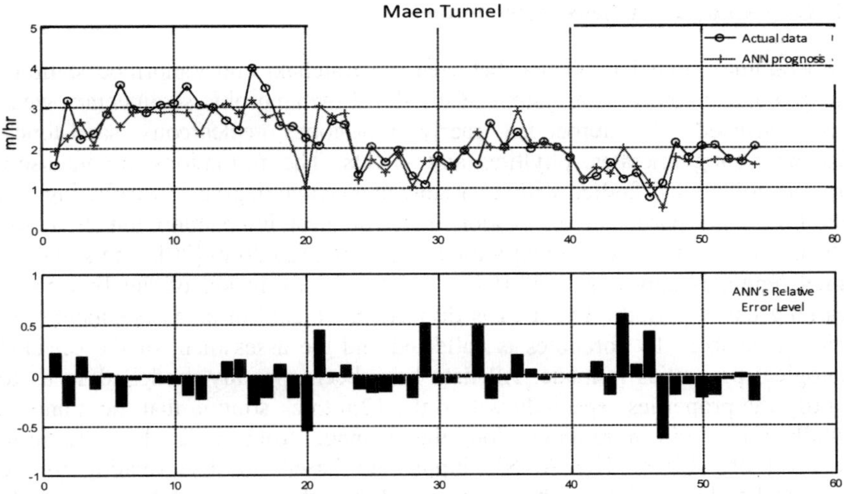

Figure 2: ANN generalisation for the complete testing subset of the Maen tunnel.

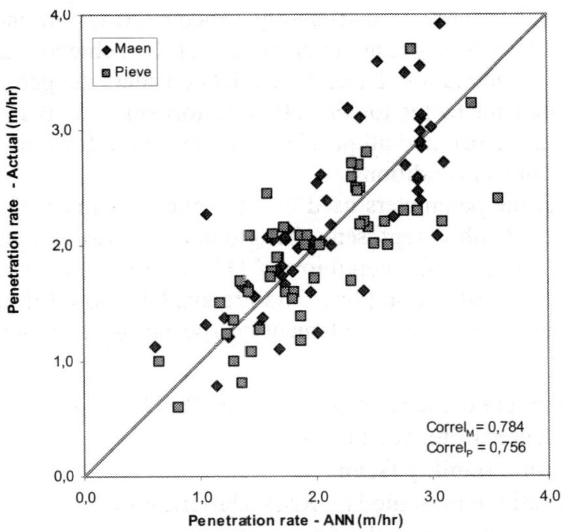

Figure 3: Scatter-plot of the measured PR values against the ANNs' predictions for the Maen and Pieve tunnels.

consequently through their respective tunnel sections. They follow the changes experienced in the actual TBM's penetration rate with satisfactory levels of accuracy and finally attain a correlation coefficient that exceeds 75% in the two examined case studies.

3.2 Case study 2 – Athens metro tunnel

The examined tunnel is located between the Katehaki and Panormou stations. The geological setting is a system of low-level metamorphic sedimentary weak rock consisted of interbedded marly limestones, calcareous sandstones, siltstones, conglomerates, phyllites and schists. The formations are intensely thrusted, folded and faulted with a variable and erratic degree of weathering and alteration. This particular excavation is the longest interstation tunnel in the Athens Metro, until now, having a total length of 1129.36 m [20]. The surveyed tunnel length is approximately 1077 m, after the exclusion of the first 53 m (learning curve period). The area is divided in 11 control areas (segments), in which, data from 16 boreholes is collected and the assessment of the selected geological properties is made. All data have been spatially modelled so as to identify the properties especially within the 12m thick stratum that the tunnel is actually being built in, ranging, along the chainage, from the level of +120m to the level of +156m. The ANN's inputs are based on data relating to the geological and geotechnical characteristics of the subsurface and the specific site conditions. Although machine characteristics (*e.g.* thrust, torque) are very important for the overall TBM performance, in the case where tunnelling is performed in soft rock or complex ground formations, the properties of the ground medium tend to be the most influential ones, as they govern the type and extend of possible failures. Subsequently, encountering ground conditions different from the TBM's working envelope, affect the achieved tunnelling rate [21] and can give rise to claims. Thus, the model considers the geological setting to be the most dominant factor for the TBM performance, as many researchers have also noted [22, 8, 10], and all possible problems and downtime are a direct effect of the geotechnical conditions.

The selection of the parameters used in the model was made having in mind their capability to credibly represent the ground behaviour, hydrogeological environment and site-specific conditions [23]. These parameters are easily collected in the site-investigation phase and are available to all design stages of the project, without the need for implementing special investigation techniques. More specifically, these are:

- Rock mass fracture degree as represented by RQD
- Weathering degree of the rock mass
- Overload factor – stability factor (N)
- Rock mass quality represented by RMR classification
- Uniaxial compressive strength of the rock
- Overburden - construction depth
- Hydrogeological conditions represented by the water-table surface relative to the tunnel depth
- Rockmass permeability

For each segment, a corresponding value for every principal parameter is taken. Allocating a representative value for the parameters is accomplished by the spatial modelling of the parameter's value and by the incorporation of statistical distribution that characterise the parameter's behaviour in each segment [30].

In the next step, the data is categorised in 4 interval scale classes, from 0 to 3, where 0 denotes the worst case and 3 the best. The limits taken in every class are representative of the specific site conditions and the machine characteristics. In the case of the Athens metro, the tunnel is constructed in relative low depth and, in general, in weak rock conditions with a double shield TBM machine. The rating of each parameter is presented in Table 5.

Table 5: Rating of the parameters.

Rockmass Fracture degree - RQD	
Value Class	*Rating*
< 10	0
10-30	1
30-60	2
> 60	3

Overload Factor (N)	
Value Class	*Rating*
> 5	0
3-5	1
1,25-3	2
< 1,25	3

UCS (MPa)	
Value Class	*Rating*
< 2	0
2-15	1
15-40	2
> 40	3

Water Table Surface (m)	
Value Class	*Rating*
> 10	0
5-10	1
0-5	2
< 0	3

Rockmass Weathering	
Value Class	*Rating*
Compl. Weath.-CW	0
High Weath.-HW	1
Med. Weath.-MW	2
SW, Fresh	3

Rock Mass Rating - RMR	
Value Class	*Rating*
< 10	0
10-30	1
30-60	2
> 60	3

Overburden (m)	
Value Class	*Rating*
< 7,5	0
7,5-12,5	1
12,5-17,5	2
> 17,5	3

Permeability (m/sec)	
Value Class	*Rating*
$< 10^{-4}$	0
10^{-4}-10^{-6}	1
10^{-6}-10^{-8}	2
$> 10^{-8}$	3

The limits of the proposed rating transforms the continuous data to a discrete probability structure, a form that is finally used as input to the model. More specifically, the data is introduced to the ANN as the expected values (EV) of the parameters (Table 6). For example, given $V_1, V_2, \ldots, V_n$ values having a respective probability of occurrence $P_1, P_2, \ldots, P_n$, the expected value of the variable X, is estimated as:

$$E[X] = EV = \sum_{i=1}^{n} P_i \cdot V_i \,, \quad \text{while,} \ \sum_{i=1}^{n} P_i = 1 \qquad (5)$$

The tunnelling advance rate (AR), recorded in each segment (Table 7), is also introduced into the ANN model. Hence, the input vector of the parameters is tallied to the output vector of the mean achieved advance rate, in each segment, expressed in m/day [21]. Note that all externally originated delays (*e.g.* strikes, maintenance, etc.) have not been taken into account.

Table 6: Expected values of the parameters in each segment.

Parameter	Seg1	Seg2	Seg3	Seg4	Seg5	Seg6	Seg7	Seg8	Seg9	Seg10	Seg11
RQD	0.13	0.88	0.90	0.64	0.72	1.37	1.62	1.26	0.55	0.66	0.74
Rockmass Weathering	2.52	2.52	2.24	1.97	1.99	1.89	1.95	1.93	1.96	1.94	1.93
Overload Factor	1.07	0.89	1.92	1.99	2.73	2.16	2.49	2.28	2.43	2.61	2.43
Rock Mass Rating	0.00	0.00	0.36	0.93	1.10	1.83	2.00	1.49	1.08	1.00	1.00
UCS	0.48	0.57	0.97	1.06	1.68	1.31	1.28	1.20	1.16	1.21	1.15
Overburden	0.42	1.00	1.17	1.97	2.86	2.35	1.16	1.45	1.13	0.99	0.88
Water Table Surface	3.00	2.32	1.71	1.00	0.23	0.02	0.94	1.40	2.17	2.40	2.75
Permeability	1.92	1.97	1.95	1.89	1.86	1.69	1.90	1.82	1.86	1.76	1.81

Table 7: Tunnelling advance rate data in each one of the control segments.

Segment	Average AR (m/day)	Max AR (m/day)	Min AR (m/day)
1	4.00	8.8	0.0
2	4.54	8.8	0.0
3	6.25	10.4	2.8
4	4.35	13.5	0.0
5	9.82	12.1	0.5
6	9.09	13.7	7.3
7	16.67	21.0	14.7
8	11.11	18.3	4.4
9	10.85	17.0	6.1
10	12.50	17.3	1.6
11	14.07	14.8	10.4

The dataset of the whole 11 segments has been divided into two subsets. The first one (*training subset - A*) is used for the ANN's training, whereas the second (*testing subset - B*) is used for assessing the model's generalisation capability. In order to ensure the ANN's performance the testing subset is consisted by the most representative segments, in terms of the achieved advance rate, namely segments no. 2, 7 and 9, as they represent the worst, the best and an average case. From the various network architectures that were examined, the ANN that was finally selected has an 8x9x4x1 topology. The mean squared error (MSE) of training approximates at $1.4\text{x}10^{-27}$ and is attained after 103 training epochs. The results generated from the trained model were very satisfactory, as the elative error (Δ) between the model outputs and the testing subset ranges in the region of 6% and 8% (Table 8).

Table 8: ANN generalisation output and actual AR data for the testing subset.

Segment	ANN generalisation results	Actual data	Relative error
2	4.854	4.54	0.0693
7	17.687	16.67	0.0610
9	9.942	10.85	-0.0837

www.witpress.com, ISSN 1743-3509 (on-line)

4 Concluding remarks

The utilization of artificial intelligence techniques, like the artificial neural networks, in TBM performance prediction can produce reliable solutions and can contribute in the efforts of their better understanding. This has been the case in the projects analysed in the paper, where the developed networks could efficiently and consistently generalise the behaviour of the three TBMs in their respective geotechnical environment.

The final remarks can be drawn:

- The use of ANN can provide an easy and user friendly modelling environment with enhanced capabilities.
- Once trained, the ANN can become an efficient tool for the prediction of the TBM's performance. It is a very flexible system and its feed with updated construction data could improve its accuracy and expand its applicability limitations.
- In terms of identification risk prone areas the use of investigation data in the ANN model could facilitate in the planning phase of tunnels, in selecting tunnel alignment, to the selection of TBM characteristics or even in selecting the most appropriate ground improvement technique.

As a final point, it should be noted that data and case records from projects already constructed could be gathered in a extensive database covering all aspects of physical, geological, geotechnical, as well as TBM and site specific characteristics. This could be an important first step to have a "universal" ANN development, which could integrate all past experience so as to generalise solutions and provide answers to all critical issues.

Acknowledgements

The author would like to thank Dr. M. Berti from the University of Bologna for providing the complete data of the Maen and Pieve tunnels and for giving his permission for their analysis in the context of this publication.

References

[1] Alber, M., Advance rates of hard rock TBM's and their Effect on Project Economics. *Tunnelling and Underground Space Technology*, **15(1)**, pp. 55–64, 2000.

[2] Barla, G., & Pelizza, S., TBM Tunnelling in difficult ground conditions. *GeoEng 2000*, Melbourne, 2000.

[3] Eisenstein, Z., Urban tunnelling challenges and progress. *ITA 25th Anniversary Commemorative Book,* 1999.

[4] Tarkoy, P.J., Predicting TBM penetration rates in selected rock types. *Proc. Ninth Canadian Rock Mechanics Symposium*, Montreal, 1973.

[5] McFeat-Smith, I., & Tarkoy, P.J., Assessment of Tunnel Boring Performance. *Tunnels and Tunnelling*, pp. 33–37, 1979.

[6] Bruland, A., Prediction model for performance and costs. *Norwegian TBM Tunnelling*, Norwegian Tunnelling Society, pp. 29–34, 1999.

[7] Sharp, W., & Ozdemir, L., Computer modelling for TBM performance prediction and optimization. *Proceedings, Int. Symp. on Mine Mechanization and Automation*, CSM/USBM, pp. 57–66, 1991.

[8] Nelson, P.P., TBM performance analysis with reference to rock properties. *Comprehensive rock engineering*. Pergamon Press, pp. 261–291, 1993.

[9] Barton, N. TBM *Tunnelling in jointed and faulted rock*, Balkema, Rotterdam, pp.173, 2000.

[10] Sapigni, M., Berti, M., Bethaz, E., Busillo, A., & Cardone, G., TBM Performance Estimation Using Rock Mass Classifications. *Int. J. Rock Mech & Min Sc.*, **39(6)**, pp. 771–788, 2002.

[11] Gong, Q.M., & Zhao. J., Influence of rock brittleness on TBM penetration rate in Singapore granite. *Tunnelling and Underground Space Technology*, **22(3)**, pp.317–324, 2007.

[12] Yagiz, S., Utilizing rock mass properties for predicting TBM performance in hard rock condition. *Tunnelling and Underground Space Technology*, **23(3)**, pp. 326–339, 2008.

[13] Bruines, P., Neuro-fuzzy modelling of TBM performance with emphasis on the penetration rate. *Memoirs of the Centre of Engineering Geology*. Delft, no 173, 1988.

[14] Alvarez Grima, M., Bruines, P.A., & Verhoef, P.N.W, Modelling tunnel boring machine performance by neuro-fuzzy methods. *Tunnelling and Underground Space Technology*, **15(3)**, pp. 259–269, 2000.

[15] Okubo, S., KFukui, K., & Chen, W., Expert system for applicability of tunnel boring machines in Japan. *Rock Mech. & Rock Eng.*, **36(4)**, pp. 305–322, 2003.

[16] Benardos, A.G., & Kaliampakos, D.C., Modelling TBM performance with artificial neural networks. *Tunnelling and Underground Space Technology*, **19(6)**, pp. 597–605, 2004.

[17] Zhao, Z., Gong Q., Zhang Y., & Zhao J., Prediction model of tunnel boring machine performance by ensemble neural networks. *Geomechanics and Geoengineering*, **2(2)**, pp. 123–128, 2007.

[18] Fausett, L., *Fundamentals of neural networks. Architectures, Algorithms and Applications*, Prentice Hall International Editions, 1994.

[19] Sietsma, J., & Dow, J.F., Creating artificial neural networks that generalize. *Neural Networks*, **4**, pp. 67–79, 1991.

[20] Attiko Metro SA. *Interstation Katehaki – Panormou: General construction report*, Attiko Metro, Athens, 1995.

[21] Deere, D.U., Adverse geology and TBM tunnelling problems. *Proc. RETS, Society of Mining Engineers*, vol. 1, pp. 574–586, 1981.

[22] Tarkoy, P.J., Tunnel boring machine performance as a function of local geology. *Bul. Assoc. Engineering Geology*, vol. xvii, no.2, pp. 41–44, 1981.

[23] Benardos, A.G. & Kaliampakos, D.C., A methodology for assessing geotechnical hazards for TBM tunnelling - illustrated by the Athens Metro. Greece, *Int. J. Rock Mech & Min Sc.*, **41(6)**, pp. 987–999, 2004.

Reinforcement fibers in concrete envelopes of underground nuclear power stations

V. Doležel[1] & P. P. Procházka[2]
[1]*University of Pardubice, Czech Republic*
[2]*Assoc. of Czech Concrete Engineers & CTU, Prague, Czech Republic*

Abstract

Underground spaces offer large areas or volumes for establishment of underground nuclear power stations, underground halls, underground deposits of nuclear waste, and underground sewerage plants, etc. The roofing of such structures requires thick walled structures, in most cases being created from fiber reinforced concretes. Additionally, standard rebars serve as a bearing reinforcement while the fibers keep off moisture, chemical gas, vapor, which can cause damaging corrosion of the rebars of various kind. The fibers serve also as defense from influence of relaxation due to change of temperature. Since the structures of this kind are of length span, the construction of them demands special treatment during the soil covering of the roof of such structures, which are here considered pelted. Moreover, very important phenomenon, creep, should be involved in the calculation as the time for building up such robust structures requires long period, during which the creep in particular parts of the structure can influence the stress state in the whole structure. The starting idea is based on creation of lathwork supporting the whole structure. On this lathwork all parts of the concrete structure will be positioned in stages, which are prescribed with respect to successive loading and optimal bearing capacity in overall structure involving successive influence of creep.

In this paper, advantages of surface nuclear power stations, underground drilled power stations and pelted nuclear power stations are discussed and for the latter fiber reinforce concrete is discussed. Some results of tests of selected fibers aiming to application if structures of pelted power stations are presented. The influence of mechanical behavior, as well as the thermal and chemical effects is shown.

Keywords: underground power stations, pelted nuclear power stations, fiber reinforcement.

WIT Transactions on the Built Environment, Vol 102, © 2008 WIT Press
www.witpress.com, ISSN 1743-3509 (on-line)
doi:10.2495/US080041

1 Introduction

It seems likely that many of the world's states will soon begin to build many nuclear power stations; some for the first time and others after ending a long-frozen program. The reasons cited centre on climate change as it is true that, once operational, nuclear reactors are largely carbon-neutral. Furthermore, they have high energy density (very high power output from a very small space) and operate continuously over lengthy periods. All they do is provide a framework in which a controlled fission reaction within its uranium fuel heats up a primary coolant (circulating water or inert gas, contained under pressure). The super-hot coolant then heats water via a heat exchanger to raise steam to drive turbines to generate electric power. Renewables have low energy densities and operate intermittently regardless of the source of energy. At present, there is no viable way to store energy produced on a large enough scale to keep power available at all times; something we have come to expect. These factors, among others, make it inevitable that many new reactors will be built.

Given that reactors will be built whether we like it or not, how can we ensure that they are as safe as possible? Mention the word 'nuclear' to most people, and words like Chernobyl, Hiroshima, missiles, nuclear waste, Windscale and Three Mile Island trip into the mind. Nuclear power has not, over the years, had a good press. Yet it could easily be made much safer.

It is necessary top take into consideration nuclear power hazards. These are well known so we'll just briefly review them. The hazards all stem from the radiation produced by the primary heat-generating fission reaction, spent fuel rods, irradiated reactor assemblies, reprocessing (if any) and the resulting radionuclides which are created in the fissioning of uranium-235 atoms. The reactor is typically sealed in a primary containment vessel with radiation shielding surrounding it. These assemblies, in turn, are usually contained in a secondary reinforced concrete building which is designed to contain radiation products in the event of an accident in which the primary containment breaks down. There was no secondary containment at Chernobyl and the results of the partial meltdown that followed the doomed 'experiment' are now grim history.

2 General strategy

The underground structures suppose to be equipped with some important accessories. One of the most essential appears to be a defense from influence of chemically aggressive gas and such other matters, from moisture and vapor, suppressing volumetric changes during curing process of concrete envelope, and diminishing of impacts of relaxation due to change of temperature. Consequently, properly prepared fiber reinforced concrete has to be used for basic parts of the structure of power station.

Usage of existing software equipment for heating balance of power cycle for resolving the heat extraction and for resolving circulation of refrigerating media (thermo hydraulics of fuel zone, residual heat extraction to the atmosphere).

Program systems based on FEM, BEM, and SHDM, which are contemporary available at University of Pardubice and Civil Engineering of the Czech Technical University will be used for comparative and combined problems. Program for calculation of average development of elastic constants of ground and terrestrial environment from deformation defined on physical model will be used for inverse analysis and interpretation.

The effective usage of mathematical models is limited by the fact that only in relatively very small field of set of tension Hooke′s linear law governs in ground or terrestrial massive. A number of experiments to express physical non-linearity have been done in several last decades. However, this effort often dash against the basic ignorance of the physical law. Unless theoretically defined physical rule of law is not entirely speculative, it is necessary to result from the laboratory results obtained on specimens or fragments of the ground, eventually from metering in situ. In the first case the results have only limited relevance; in the second case some required tests are only very hardly viable and very capital-intensive.

Usage of the experimental method of physical simulation can entirely eliminate above-mentioned faults and problems with suitable strategy of experiments. Further this experimental method allows us progressive survey of transformation of particular substances to the limit of failure, what makes also combined simulation with usage of suitable formulated mathematical and physical models very attractive.

General process in combined analysis of mechanics of ground and soil problems involves:

1. Construction of geological profiles of the ground complex
2. Dividing planes will be marked out (bedding, foliation, fissures, cracks and dislocations), which notably influence resolving the task. Eventual control of the importance of particular dividing planes will be made on the physical models.
3. Relation between stresses, strains, deformation, speed of deformation and time by mechanical tests of homogenous or quasi-homogenous parts of the grounds are to be determined. The tests will be made in order to be able to appoint beginning and process of the grounds dilatation. It is necessary to choose appropriate development of the stresses before the failure of the system.
4. Mechanical characters of the dividing planes (deformation characteristics, cohesion, friction, dilation) and filling substances by measuring in the terrain are to be discovered.
5. A model on the physical similarity principles will be created to respond the ground complex.
6. Input parameters of the ground complex for mathematical solution of the problem will be prepared from the results of experiments on physical models.

3 Pullout problem of a steel FRC beam

The pullout problem has frequently been solved in a problem of cracking of composite structures of several sorts. Since large span structures are considered here, locally large deformation pull out problem should be taken into account.

In this study of a lag model of the system concrete matrix - steel fiber is used. A finer mesh of the finite elements is necessary for the stability of an iterative process particularly in the neighborhood of expected nonlinear behavior on the contact (interface between fibers and matrix).

The debonding or “slipping" (jumps of responding points of the matrix and fibers boundaries in the tangential direction to the interface) can be caused due to several physical models. First, no tension tractions on the interface may occur. Also, some friction law (such as Coulomb law, Mohr-Coulomb law, shear bond strength are often introduced and debond parameters are computed. Since no fiber gripping force from the matrix is expected, the general Mohr-Coulomb friction law is restricted to the exclusion of shear stresses exceeding the shear strength in the tangential direction and the tensile stress exceeding the tension strength is also excluded along the interface between the fiber and the matrix.

The practical reason for applying large displacement theory shows the following Fig. 1. From this picture it is seen that the fibers are pulled out of the concrete and the behavior during the pulling process is much more complicated than when starting with small displacement theory. This is also the motivation of this study.

Figure 1: Detail of the cracked zone.

In the cracked zone it is seen that the fibers have been pulled out of the matrix in such a way that their shape is almost hold and the process of pullout is rather more complicated then it was considered before. Curvilinear Dramix type fibers have been used. Similar situations can occur when straight fibers non-symmetrically positioned are applied, although the curvilinear are more sensitive of the way of loading.

Consider a contact problem of two bodies. The first body (fiber) occupies in undeformed state the domain Ω' with the interfacial boundary Γ' (generally partly rectangular parallelepipeds with three connected parts, two are parallel and the second creates slope of the fiber) and the second body (concrete matrix) occupies the domain Ω'' with the boundary Γ'', (parallelepiped), the fiber in which is embedded.

There is no external load considered but the pullout force at the face of the rectangular parallelepiped. The bodies are situated in Cartesian coordinate system *Oxyz*; the axis *x* is introduced in the radial direction of the lag. Generally, it will be assumed that the tension exceeding the tensile strength in the normal direction to the interface fiber-matrix Γ is not admitted, so that both bodies may disconnect (they mutually debond) in certain region of the common boundary Γ, which is a part of Γ' and Γ''. The limit strength contact condition will be taken into account.

The load is considered in the following way. The pullout traction *F* is applied at the face of the fiber in the axial direction, and prescribed by a starting constant value. Then the value is successively increased. There is symmetry about the vertical plane containing the horizontal axis *x* regarding both geometry and loading, so that only one half of the unit cell may be solved.

Displacements are described by the vector function $\boldsymbol{u} = \{u, v, w\}$ of the variable $\boldsymbol{x} = \{x, y, z\}$. Denote $\Omega \equiv \Omega' \cup \Omega''$. The restriction of any function to Ω', or Ω'' is denoted, respectively, by one prime or by two primes, e.g., $\boldsymbol{u} / \Omega' = \boldsymbol{u}'$ and $\boldsymbol{u} / \Omega'' = \boldsymbol{u}''$. On the boundary displacements and tractions are prescribed in such a manner that a periodicity is assumed at all sides of a "unit cell".

Denote the set of admissible displacements $\boldsymbol{u}$ on Ω satisfying the essential boundary conditions by *V*. The values and the first derivatives are quadratically integrable.

Consider Hooke's law in the form:

$$\varepsilon'_{ij}(u') = L'_{ijkl}\sigma'_{kl} \qquad \varepsilon''_{ij}(u'') = L''_{ijkl}\sigma''_{kl} \tag{1}$$

where L' and L'' are the material stiffness matrices of the fiber and the concrete, respectively. The overall stiffness matrices $\boldsymbol{K}$ relate nodal forces and nodal displacements as:

$$\boldsymbol{K}'\boldsymbol{u}' = \boldsymbol{p}' \qquad \boldsymbol{K}''\boldsymbol{u}'' = \boldsymbol{p}'' \tag{2}$$

Note that while the two primed stiffness matrix is regular one primed stiffness matrix is singular and will need certain necessary improvements after introducing the overall boundary conditions. This will be done in what follows.

Assuming "large deformation" theory, it may be satisfactory to formulate the essential boundary conditions on the contact boundary in nodal points of the elements (not necessary mutually adjacent) as follows (Signorini's conditions):

$$[u]_n = u'_n - u''_n < 0 \qquad \text{a.e. on } \Gamma \tag{3}$$

Denote

$$H \equiv \{\boldsymbol{u} \in V;\ [u]_n < 0 \quad \text{a.e. on } \Gamma\ \}$$

The set H is a cone of admissible nodal displacements with respect to the essential boundary and contact conditions, V are displacements from the space of continuous functions.

Suppose that we disconnect both bodies under consideration, but keep the stress and deformation state in them "frozen". Then the vector of nodal contact tractions $\boldsymbol{p} = \{p_x, p_z, p_y\}$ must be introduced and their equilibrium (action and reaction law) says that:

$$\boldsymbol{p} = \boldsymbol{p}' + \boldsymbol{p}'' = \boldsymbol{0} \tag{4}$$

The shear bond strength condition with exclusion of tension read as:

$$p_n - p_n^+ \leq 0, \quad [u_n] < 0\,, \quad (p_n - p_n^+)[u_n] = 0 \tag{5}$$

$$C = k(p_n - p_n^+)c\,,\ |\,p_t\,| - C \leq 0,\ \ |[u_t]| > 0\,,\ \ (|\,p_t\,| - C)\,|[u_t]| = 0\,,$$

where C is the cohesion (shear strength), k is the modified Heaviside function being equal to one for negative arguments and zero otherwise, $[u_n] = u_n^c - u_n^s$, $[u_t] = u_t^c - u_t^s$, and u_n^c, u_n^s are displacements normal to the steel – concrete boundary along the interface with respect to the concrete and the steel, respectively, and u_t^c, u_t^s are tangential displacements along the interface with respect to the concrete and the steel, respectively, p_n^+ is the tensile strength. Variational principle then leads to the following definition: Find the minimum displacement vector $\boldsymbol{u}$ and the maximum traction vector $\boldsymbol{p} = \{p_n, p_t\}$ for the functional of entire energy E_{ent}:

$$E_{ent} = \frac{1}{2}\Big(\int_{\Omega'} \boldsymbol{\sigma}^{\mathrm{T}} \boldsymbol{\varepsilon}\, \mathrm{d}\Omega' + \int_{\Omega''} \boldsymbol{\sigma}^{\mathrm{T}} \boldsymbol{\varepsilon}\, \mathrm{d}\Omega''\Big) - \int_{\partial\Omega''} \boldsymbol{p}^{\mathrm{T}} \boldsymbol{u}\, \mathrm{d}\partial\Omega'' + \tag{6}$$

$$+ \int_{\Gamma} (|\,p_t\,| - C)\,|[u_t]|\, \mathrm{d}\Gamma + \int_{\Gamma} (p_n - p_n^+)[u_n]\, \mathrm{d}\Gamma$$

where subscripts in denotation of domains mean c = concrete, s = steel, $\boldsymbol{\sigma}$ and $\boldsymbol{\varepsilon}$ are respectively the stresses and the strains in both bodies, Γ is the interface, $\partial\Omega''$ is the external boundary of the representative element, T means transposition. It is usable to introduce the mixed formulation for the finite element method:

Let us write the total energy J of both bodies assuming them separately:

$$J(\boldsymbol{u}, \boldsymbol{p}) = \Pi(\boldsymbol{u}) - I(\boldsymbol{u}, \boldsymbol{p}), \tag{7}$$

where

$$\Pi(\boldsymbol{u}) = \frac{1}{2} a(\boldsymbol{u}, \boldsymbol{u}) - \int_{\partial\Omega''} (\boldsymbol{p}'')^{\mathrm{T}} \boldsymbol{u}''\, \mathrm{d}\partial\Omega''\,,$$

$$I(\boldsymbol{p}, \boldsymbol{u}) = \int_{\Gamma} (|\,p_t\,| - C)\,|[u_t]|\, \mathrm{d}\Gamma + \int_{\Gamma} (p_n - p_n^+)[u_n]\, \mathrm{d}\Gamma\,, \tag{8}$$

$$a(\boldsymbol{u},\boldsymbol{u}) = \int_{\Omega'} \boldsymbol{\sigma}^{\mathrm{T}} \varepsilon \, \mathrm{d}\Omega' + \int_{\Omega''} \boldsymbol{\sigma}^{\mathrm{T}} \varepsilon \, \mathrm{d}\Omega''$$

The first stiffness matrix in (2) appears to be singular, while the second is regular, as the second matrix is created under assumption of periodic conditions. For the first matrix in order to be regularized has to be connected with the first by introducing some relation to the second matrix by connection of springs, for example. This trick enables us to connect both substructures (fibers and matrix) in a physically reasonable way.

From the above consideration it follows that the contact representation by spring stiffnesses k_n (normal direction) and k_t (tangential direction) can be characterized (normal and tangential directions are taken with respect to the surface of the fiber) and penalty like formulation can be considered in both cases: small deformation and large displacement theories. If the spring stiffness is high, the bond of fibers to matrix is defined, if some of contact conditions are violated, the stiffness lowers its value. The impact of this formulation ensures always the solution, if the fiber is not disconnected from the matrix in all nodal points. In this sense the interfacial relations can be written as:

$$p_n = k_n[u_n] \qquad\qquad p_t = k_t[u_t] \tag{9}$$

and the interfacial energy I in (8) is written as:

$$I(\boldsymbol{p},\boldsymbol{u}) = \int_\Gamma k_t[u_t]^2 \,\mathrm{d}\Gamma + \int_\Gamma k_n[u_n]^2 \,\mathrm{d}\Gamma - \int_\Gamma C\,|[u_t]|\,\mathrm{d}\Gamma - \int_\Gamma p_n^+[u_n]\,\mathrm{d}\Gamma \tag{10}$$

where spring stiffnesses k play the role of penalties and the last terms in (10) represent the peak (strength) energies for both normal and tangential directions.

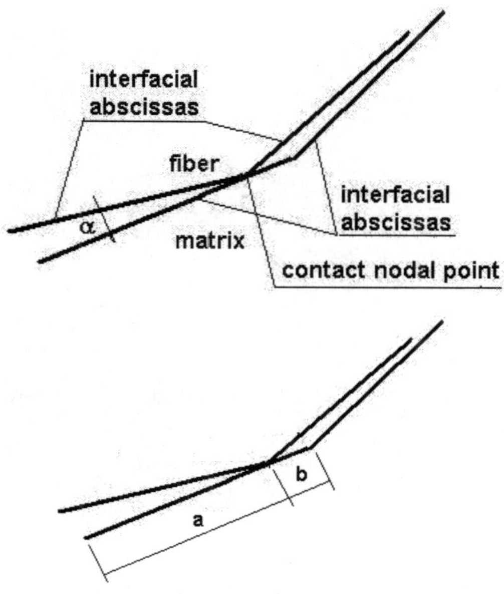

Figure 2: Calculation of forces after dissipation of interfacial energy.

4 Algorithm

In the case of large displacements two possibilities have to be distinguished. The first consists in the situation which appears after disconnection of fibers from matrix. Then the interfacial energy disappears and if no new fiber-matrix contacts occur, the crack remains open. If the new touch of fibers with matrix would be found, see Fig. 2, new contact forces are generated according to "spring rule". The forces are divided into nodal points regularly, i.e., if the force is, say, F, then to the node near a we get F b/(a+b) and to the node near b the force F a/(a+b) is transmitted. The similar formulation (7) takes place and the algorithm continues in the standard way.

If the angle α is equal to zero, the accumulated energy due to shear and normal forces does not disappeared and has to be taken into consideration. The new position of the nodal points has to be considered and additional energies are generated from the new constellation of the nodes. The energy functionals differ from the starting one, (7), by additional terms which are based again on "spring rule".

5 Experiments with fibers

In order to compare polypropylene fibers (PFRC) and modified polypropylene fibers (MPFRC) and steel fiber reinforced concrete (SFRC), four point bending tests have been carried out. In Fig. 3, results from four tests have been conducted

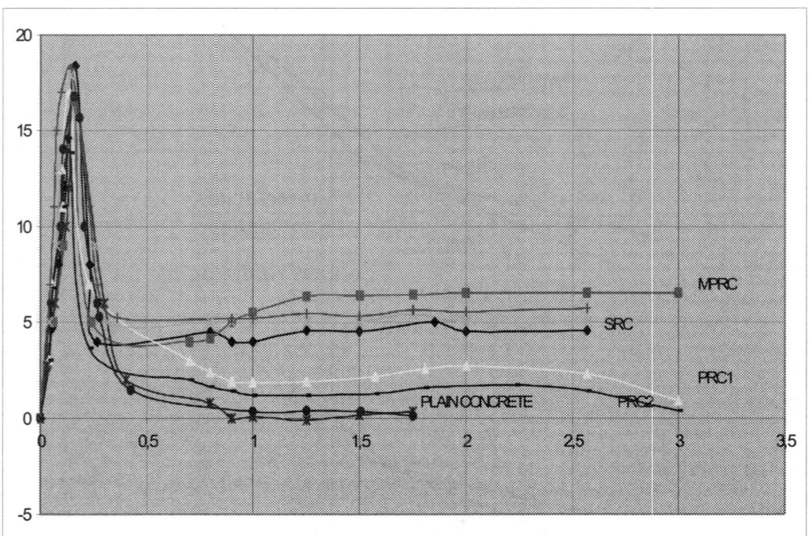

Figure 3: Comparison of results of all tests.

on 400 mm spanned beam, 100 x 100 mm^2 was the cross-section. In the picture graph force and displacement is depicted. From Fig. 3 one can see the results from three tests with polypropylene fibers, one with steel fibers and one with modified polypropylene fiber. From the picture it is seen that there is much higher residual stress by MPRC and even local hardening in the region of softening appears. The hardening seems to be very steep. In the case of polypropylene fibers slight hardening is seen, too, but not that emphatic. In Fig. 4 a similar tests has been carried out for SFRC with obviously harder stiffness. The tests show that the results are much closed one to each other.

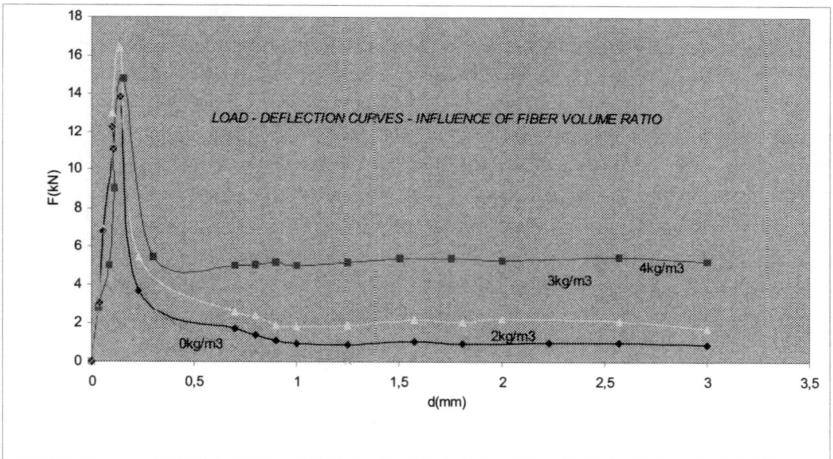

Figure 4: Influence of fiber volume ratio.

6 Conclusions

In this paper application of various fibers in large span structures is discussed. Particularly, envelopes in underground nuclear power stations are such where fiber reinforced concrete even has to be used for different reasons, which are mentioned in the above text. In case large span structures have to be the bearing element, pull of fibers out of concrete matrix cannot be simplified in calculation starting with small deformation theory. This is why large deformation theory is mentioned here and briefly described. Results from calculations are shown for various fibers with various shapes.

Acknowledgements

Financial support of GAČR, project number 103/08/1197 is appreciated. The research has also been supported by a grant of Ministry of Education of the Czech Republic number MSM6840770001.

The hydrogeological problems of disused mines in Olgiate Molgora (LC)

L. Longoni & M. Papini
Politecnico di Milano, Italy

Abstract

The object of this work is to evaluate more effective methodological approaches to face problems concerning areas affected by the presence of disused mining sites. Some years after the closure of mines, various problems as a consequence of disuse have come to light, and also, possibilities to use this resource with touristic and cultural aims have often been ignored. From the moment they have been closed mines can cause several problems, both for safety of the mine and of the neighbouring areas (downfalls, collapse of the vault, variations in the underground circulation, etc.) so it is necessary to evaluate more suitable techniques in order to maintain mines and prevent risks related to their presence. In order to deepen this subject we studied with particular detail a pilot area representative of the described problems: the Pelucchi Mine sited in Olgiate Molgora (LC). Several problems occurred in this abandoned area: the most important being that galleries have been troubled by water infiltration phenomena, starting from the lower part of the cavity. Previously the water was pumped away by the water pipeline until the water level was within acceptable limits, but when the extraction ceased water filled the cavities and saturated all materials around. This event caused a progressive alteration of the modalities of the underground water circulation, which in the future could cause instability phenomena on the surface and in the area near to the caves. In order to check the stability condition of these mines several analyses, either direct or indirect, have been made. Indirect analysis was made with the use of Satellite Radar Interferometry, to make a complete monitoring of surface movements (subsidence due to the collapse of caves), useful for the mine collapsed forecasting.

Finally a phenomenological model has been made to define risk scenarios, necessary for Civil Protection aims and for the instability phenomena analysis.

Keywords: water circulation, mines, risk.

WIT Transactions on the Built Environment, Vol 102, © 2008 WIT Press
www.witpress.com, ISSN 1743-3509 (on-line)
doi:10.2495/US080051

1 Introduction

The closure of a mine triggers many problems regarding safety and risks. In fact the closure of a mine implies the ceasing of mining activities, and consequently of all the maintenance activities, in particular dewatering by pumping: consequently the water level increases and causes different problems for the stability of underground spaces. The case described in this paper deals with dismiss of Pelucchi mine located in Olgiate Molgora, near Lecco. The Pelucchi mine was one of the biggest mines in the Lecco zone for the extraction of cement marls; production ceased in 1954 and since then the mine has been under maintenance. The mines have been used for a long time as a water reservoir for the water pipeline. When the pumping ceased, all seven levels of the mine have been filled with water; after this, all water storage activities were definitively abandoned. This site has been analyzed as a guideline for the use of different kinds of monitoring systems, either direct (like geomechanical and hydrogeological surveys) or indirect (like Satellite Radar Interferometry). Every method has been applied and the results have been compared in order to forecast instability phenomena. Moreover, a phenomenological model has been defined to investigate all possible risk scenarios that could occur in and outside the mine.

2 Geological and geomorphologic context

Olgiate is in North of Italy, 20 km south of Lecco and few kilometres from Adda river. The area is characterized by sedimentary sequence of cretacic rocks, referring to Scaglia and Bergamo Flysch Formations, which locally show limestone-marl levels, even with thickness of ten meters. Quaternary sediments (colluvial and alluvial deposits) are diffused and continuous, covering most parts of the surface rock, although there's no difficulty in understanding underground geo-structural characteristics. Caves develop along the reversed side of Lissolo Synclinal, with South alignment and having axial plane with WNW–ESE direction. The most important tectonic features that interest Pelucchi mine are two sub-vertical faults with N–S direction.

2.1 Map of the mine

In order to apply all geomechanical classifications the mine has been divided in to different tracts with homogeneous mechanical characteristics. It has been easy to divide these mines by the structural conditions of the rock mass and by the reduced presence of sets of joints (three sets of joints in the marls and one set – bedding – in the flysch). Homogeneous tracts that resulted from the structural analysis have been called: North Entry (INNO), South Entry (INSU), Pelucchi Gallery (PELU), Fault (FAGL), Piave Gallery (PIAV), West Piave Gallery (PIOV), Final Buttero Gallery (BUFI), West Buttero Gallery (BUOV) and East Buttero Gallery (BUES). In the map below (fig. 1) it is possible to see all these tracts with their position in the mine.

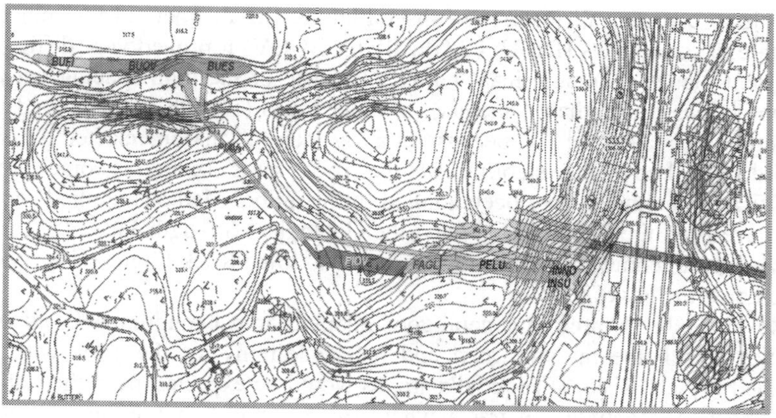

Figure 1: Tracts of the Pelucchi Mine.

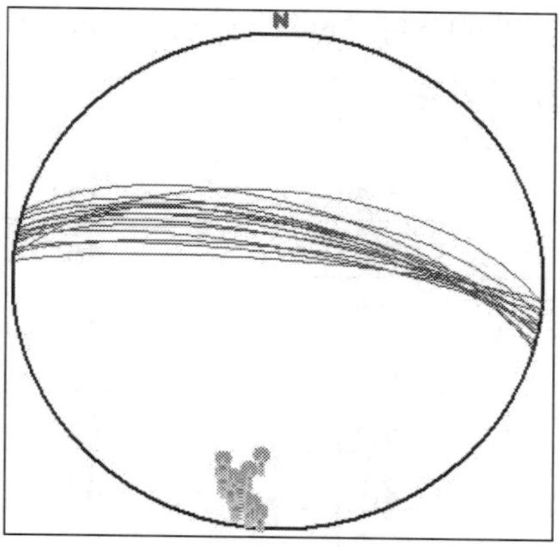

Figure 2: Example of stereographic projection (Schmidt net).

2.2 Geological survey

First, the direction of strata and joints presented in the mine has been determined. It has been possible to define geometries of discontinuities with respect to the direction of the cave (East–West direction), obtaining a good base for an evaluation of its stability. The section made by massive marl has bedding strata with dip of 5° and inclination of 60° and two joints sets, K1 with dip of 270° and inclination of 75° and K2 with dip of 15° and inclination of 35°. Spacing shows very high values in all three sets (S0, K1, K2), generally of 3–4 meters, and without any filling material (almost all the cracks are closed).The part made by thinly stratified flysch shows a dip direction similar to the marl one, with spacings of about 5cm filled by clay. All data obtained by the survey of all sets have been mapped with stereographic projections in a Schmidt net (as it is possible to see in the example of fig. 2). It is necessary to say that it's not been possible to consider all features of discontinuities in all tracts in the mine: for example in some tracts (INSU, FAGL, BUFI, BUOV) it proved very difficult to take joint planes in a significant number because of the high spacing between discontinuities (more than 4 meters); in other tracts (e.g. PIOV) it proved very dangerous to make a survey, because of a big rock fall that intersected the tract. In order to know the stability of the rock mass several geomechanical classification methods have been applied in this mine. Methods applied for this case have been the Q-system [7] and the Geological Strength Index [8] for the knowledge of geomechanical quality and the Modified Rock Engineering System [9] to know the Risk Index.

Results (table 1) present general good geomechanical conditions for all tracts associated with the massive marl and bad conditions for the parts with thin stratified Flysch. Risks of rock falls are, as would be expected from the stability assessment, concentrated in tracts where there are Flysch lithology outcrops, underlining that massive marl is stable.

Table 1: Values of results for Q-system, GSI and MRI.

	Q-system values	GSI values	MRI values
INSU	1.5	32	22.2
INNO	7.5	45	22
PELU	10	60	23.4
FAGL	6.6	52	26.4
PIAV	0.3	22	61.2
PIOV	2.2	37	24.1
BUFIN	2.2	50	20
BUOV	20	62	20.9
BUES	61	80	20.9

2.3 Hydrogeological survey

Direct analysis was also used to study the hydrogeological system of Pelucchi area; it is possible to find two sub-systems: a superficial alluvial system and a complex underground system that links tunnels with external water wells. In general, the water level inside Pelucchi mine is correlated with atmospheric water phenomena. From previous data it is reported that after long rain periods the water level rises to the first level of galleries and overrides it. An important event is the December 2002 one, when after a 10-day rainy period the water level rose to 20cm above the first level ground, flowing out of the principal access to the mine and flooding the road. In order to verify the correlation between rains and water level and then for monitoring the water level in the Pelucchi tunnels, two pressure sensors (designed for submersible measurement) were installed in

November 2006. The first one (sensor A) has been installed in the well which links first level with other levels in the Pelucchi part, while the second one (sensor B) has been installed in the well that links three levels of the Buttero side of the mine.

In fig. 3 variations of water level in the period between November 2006 and March 2007 are represented. In relation to sensor A only small variations in water level were observed until 3-7-2007, probably due to a rise in the flux from Piave linking tunnel. After this date, the position of sensor was changed, and until 1-6-2007 it revealed a lowering (135 cm) with respect to the first measured level, while in 6-7-2007 it revealed a rise of 88cm. The comparison between sensors A and B reveals that there are not any common features in water level movements: the only matches available are raising period ones, when (2–8 May 2007 and from June 2007) water level had a rise in both parts of the mine. However, it is important to note that the entity of rising was very different between A and B points (some cm for sensor B and 150 cm for sensor A). A conclusion would be that in the two examined tunnels in and out water flows are very different.

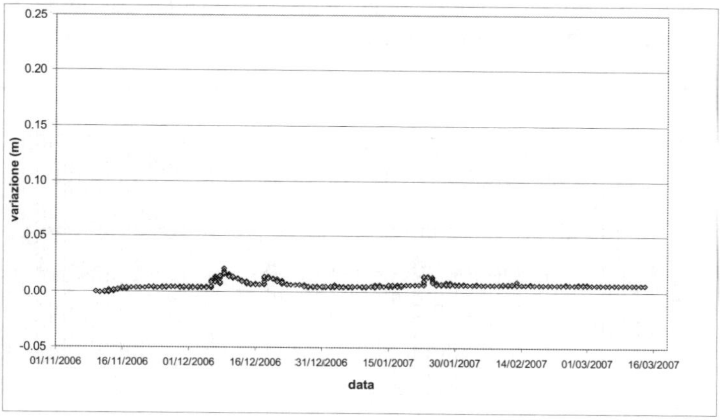

Figure 3: Water level flows in Pelucchi tunnel (sensor A) in the period 1-11-2006 – 14-3-2007.

2.4 Indirect analysis

In order to evaluate geological risk connected to the presence of hydrogeological diseases, ordinary field surveys have been integrated with the Permanent Scatters (PS) Technique, which is largely used for the monitoring of subsidence or uplift. PS technique analyzes thousands of square kilometers of territory within an extremely short time scale. The Permanent Scatters Technique is a tool that detects, measures and monitors ground movement, using satellite SAR (Synthetic Aperture Radar) data. The PS technique takes conventional differential InSAR a step forward, by identifying single benchmarks, often referred to as permanent scatters and reconstructing their displacement history.

PS are radar targets that are located across the earth's surface and can be monitored by satellites. It is possible to detect and measure millimetre variations in the sensor- target distance, over time. PS correspond to objects on man-made structures (bridges, dams, buildings, etc.) as well as to stable natural reflectors. This technique has been applied in the mine site to monitor ground movements. As it is possible to see, the PS in the area correspond to buildings; the presence of vegetation near the mines does not permit any other reflectors. It has been possible to estimate the displacement rate for which the precision can be as good as 0.1 mm/year and reconstruct the displacement history of the PS.

Analyzing track nr.251 no movement has been found (fig. 4). Other data (track 208 and 480) show that the area is subjected to uplift; track 480 instead highlights subsidence. The different movements found in another two tracks allow saying that for the Olgiate area the technique of PS doesn't provide necessary data to understand subsidence or uplift phenomena. It is also possible that no important movement has been detected, for the absence of radar targets with PS characteristics directly on the surface correspond to underground presence of caves, being all the area covered by vegetation. In fact houses sited on the north part of the area are distant from the mine and consequently it proves very difficult to correlate those points with hydrogeological and structural changes induced by the presence of mines.

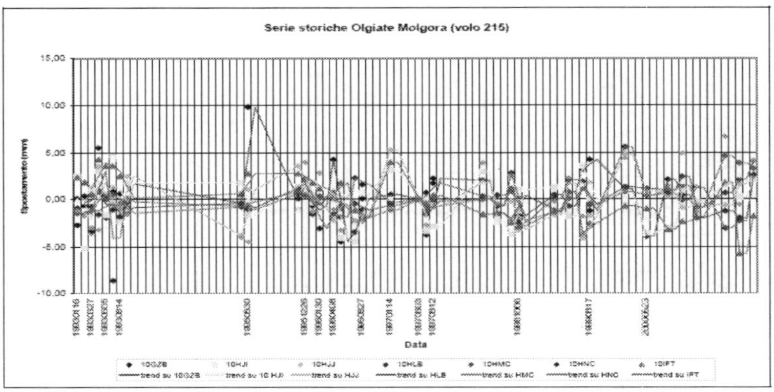

Figure 4: Historical series for Pelucchi galleries from track 215.

3 Phenomenological models

A phenomenological model has been made to define risk scenarios, necessary for Civil Protection aims and for the instability phenomena analysis. Five scenarios have been reconstructed.

Scenario 1

The first level, that is the only accessible one in Pelucchi mines, develops inside a bank of marls that is followed by a highly stratified flysch (filled with clay). In particular the first level develops through compact marls and flysch. There is one outcrop on the left part of the gallery, characterized by the presence of thin-stratified marl strata interlayered with centimetric levels of clay. In the part characterized by the presence of massive marl there is a low probability of rockfalls, while in the part of the mine characterized by the presence of flysch the probability of rockfalls is more than 50%. This fact is demonstrated by previous collapses inside caves, as is possible to see in fig. 5. So this scenario seems to be probable, and it is possible for another collapse to occur near the roof or the left wall. However, no consequence, outside of the mine, is to be considered by such collapses in the flysch part; this scenario should be considered only if, for example, these tunnels were to be used for touristic aims.

Scenarios 2 and 3: sudden collapses of sects that divide caves from the surface and sudden and extended collapses of sects and walls of tunnels

These scenarios have been set under the same title because they both address various kinds of collapses inside the caves. Such collapses can be due to the same causes, and the part of the mine where they take place would determine several effects on the stability. Factors that could cause these scenarios are similar to the ones in case 1: fracturation of rock mass, presence of clay interlayered with marls. So the difference between these two cases and case 1 is the dimension of the phenomenon: case 2 and 3 contemplate the collapse of considerable parts of caves.

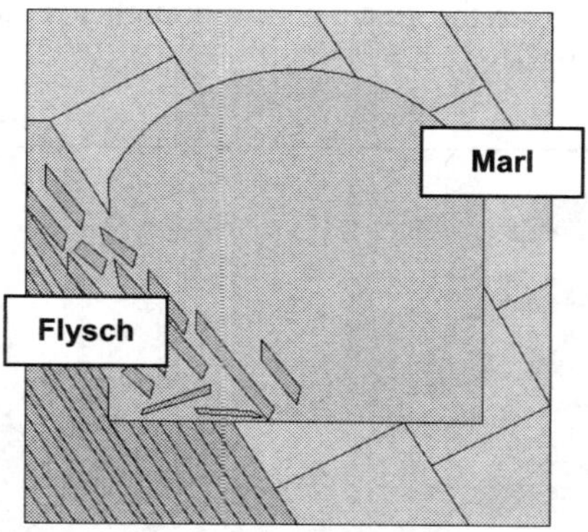

Figure 5: Scenery 1: Rockfall on flysch part (brown).

In particular, scenario 2 contemplates collapses that interest the surface (roof of level 1), while scenario 3 contemplates underground levels.

Possible consequences of both scenarios would be:

- Sudden water flow out of tunnels along the slope, with possible flooding of the inhabited area (fig. 6);
- Sudden pumping of water in second level along the slope, with possible flooding of the inhabited area. (fig. 7).

Geomechanical survey suggests that only where flysch constitutes most part of the cave can collapses of such entity occur: in fact in these tracts several collapses have already happened without any consequence.

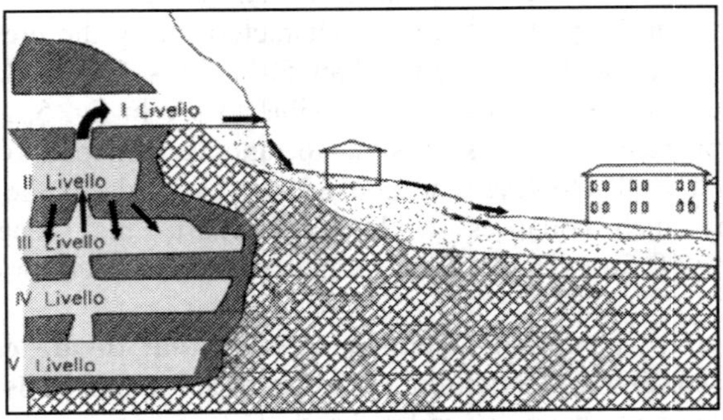

Figure 6: Scenario 2: Waterflow out of galleries.

The probability of such an event happening of could be known after an assessment of the lower levels, which can't be investigated with a direct survey, as they are filled with water. It is difficult to know the position of flysch in the underground levels because, as was said above, they are filled with water and it is very difficult to make a direct survey. So it is impossible to evaluate exactly what is the percentage of risk of underground collapse. However, the risk of a collapse of the roof over the first level, although it is a remote possibility, is always to be considered.

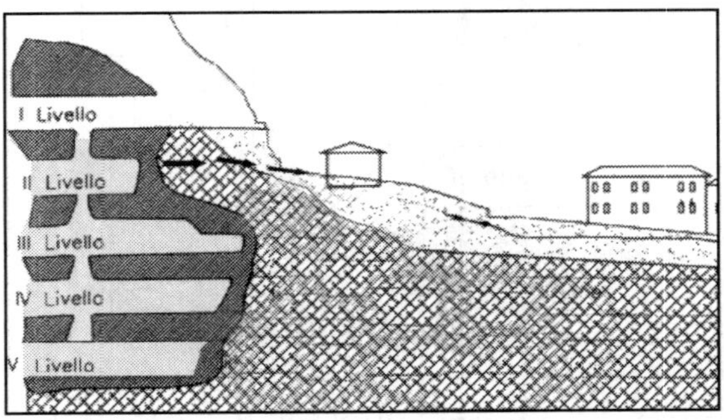

Figure 7: Scenario 3: Pumping of water present in galleries, with possible flooding.

Scenario 4: flood on the slope

This scenario could be caused by the infiltration of water inside the fractured rock mass and inside glacial and fluvioglacial overlaying deposits. Such infiltrations are made possible by the fractured rocks, caused also by alteration (considering the low presence of surface sediments).

A possible consequence of infiltration of water should be a flood that would reach the basements of houses placed on the slope under the mine. This has already happened in the past, where the basements of houses in Valicelli (an Olgiate Molgora quarter) required draining works in order to bring water out of the civil structures.

As these past events demonstrate, this scenario seems to be very realistic, so it would be necessary to drain all the area around the mines. The probability of this scenario happening is high.

Scenario 5: rising water level after long period of rainfalls
This scenario could be caused by progressive saturation of the lower levels of tunnels, until they reach the surface level. The hydrogeological statement of the area near mines suggests that after intense and prolonged rain the water level can easily reach the first level. The first consequence can be water flow out of tunnels, and then on the slope under mines (fig. 8). Even this event is very probable, already happening in 2002, when water levels reached the road that passes over the entrance of the mine.

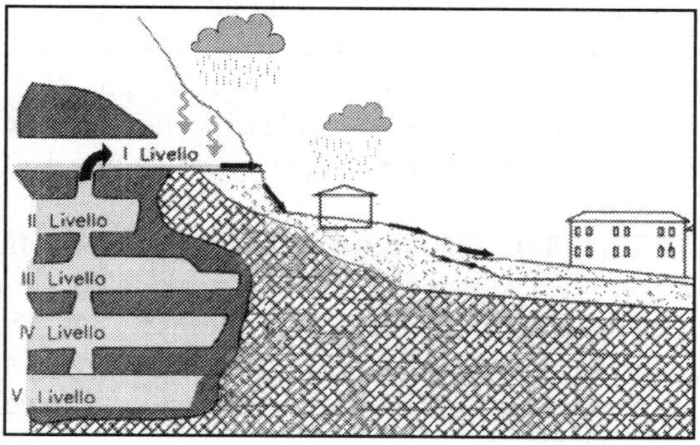

Figure 8: Scenario 5 – Flooding from entrance to the slope under the mine.

4 Conclusion

Geological and hydrogeological surveys proved to be helpful to determine the structural settlement of mines and water circulation in and outside mines. Knowing percentage of stability inside caves and the way water circulates proved to be very important to define a risk analysis, and after that to build up various possible scenarios that can occur, in order to create a phenomenological model, necessary to plan an intervention model for civil protection field works.

Moreover, the PS technique proved to be a good *remote sensing* instrument, thanks to its high reliability in identifying and monitoring ground deformations also (and first of all) in a large scale. However, note that for good results it is necessary to know very well the geological settlement of the area that is to be surveyed. In fact only with appropriate studies it is possible to know whether PS will have good significance relative to a considered phenomenon and if they will give information useful for major comprehension of the case studied.

It is important to say that morphology of the area, presence of vegetation and position of human settlement, create a non-homogeneous distribution of PS points in the area. So in cases like this one it can prove very difficult to study natural phenomena and eventual movements in the area. As an obvious consequence, in critical areas similar to this one, and for other studies in this field, it is suggested to place artificial reflectors where it is not possible to find any other pre-existent fixed point, in order to permit a complete survey of the area (especially of critical points).

References

[1] Comune di Olgiate Molgora (2005): "Labirinti sommersi: la cementeria del Fabbricone e le gallerie Pelucchi: un caso di archeologia industriale ad Olgiate Molgora". A cura di Lorenzo Brusetti, Massimo Cogliati e del Gruppo Sommozzatori Alme – Olgiate Molgora. – pp. 85 + 3 tav

[2] Penati M., Lanfranconi P. (1999): "Studio idrogeologico per lo sfruttamento delle acque presenti nelle gallerie della ex-miniera Pelucchi nel Comune di Olgiate Molgora (LC)" STUDIO GEOPLANET – pp. 20 + 5 tav.

[3] Bieniawski Z.T. (1976) - Rock mass classification in rock engineering. In: Bieniawski (ed.), Exploration for rock engineering procs. Of the Cape Town symp., 1, 97-106.

[4] Bieniawski Z.T. (1989) - Engineering rock mass classifications. Wiley Publ.

[5] Hack R. (1997) - Rock mass strength by rock mass classification. South African Rock Eng. Congr. (SARES), Johannesburg, 346-356.

[6] Papini M., Granito A. e Scesi L. (1998) - Caratterizzazione geomeccanica degli ammassi rocciosi: un approccio statistico. Quaderni di Geologia Applicata, 5(2), 41-47.

[7] Barton N., Lien R., Lunde J. (1974) – Engineering classification of rock masses fort he design of tunnel support. Rock mechanics 6 (4)

[8] Hoek E., Marinos P., Benissi M. (1998) – Applicability of the geological strength index (GSI) classification for very weak and sheared rock masses. The case of the Athens Schist formation. Bulletin of Engineering Geology and the Environment, 57

[9] Papini M., Scesi L., Gattinoni P. (2002) – Valutazione del rischio geologico nelle gallerie minerarie:applicazione del Modified Rock Engineering System. GEAM 1, 2002, 89-96

Management of complex underground construction projects

M. Leijten
Delft Centre for Sustainable Urban Areas,
Faculty of Technology, Policy and Management,
Delft University of Technology, The Netherlands

Abstract

The complexity of underground projects confronts many project clients with serious manageability problems. This complexity is the basis for this article, which analyzes the considerations that should be made when setting up a project organization in order to keep the process of implementation manageable. Project organization is analyzed closely, with a focus on the interface between project managers and functional managers and the information asymmetry between them. Some real-life examples will show that dealing with uncertainty and thereby increasing manageability are seldom accomplished by increasing available information, but rather by reconsidering the project organization.

Keywords: project management, multi-actor systems, information processing.

1 Introduction

Underground construction projects, particularly those in urban areas, ostensibly experience more significant problems than other infrastructure engineering projects. In many underground projects, technology is more challenging than in other infrastructure projects, increasing the chance of failure. Moreover, in underground projects the consequence of eventual failure is often more profound than in other projects, increasing the overall consequences of failure. As a result, underground projects have a reputation for being risky because they are susceptible to implementation problems. Society's growing demand for space, however, makes us heavily dependent on underground space to satisfy contemporary spatial claims.

WIT Transactions on the Built Environment, Vol 102, © 2008 WIT Press
www.witpress.com, ISSN 1743-3509 (on-line)
doi:10.2495/US080061

High levels of risk make underground projects more difficult to manage than many other construction projects. This article attempts to dig more deeply into the manageability of the complexity of these projects. This should provide ideas about how to set up a project organization to best handle the efforts this requires. To do so, the complexity of underground projects will first be analyzed in section 2. In section 3 the main manageability problem of organizations dealing with this complexity is identified. Subsequently, details from some real projects will show the virtues and drawbacks of some types of project organizations in the face of complexity, particularly in terms of their ability to handle manageability problems. Section 5 will briefly present conclusions about considerations that should be taken into account when designing an organization for an underground project.

2 Complexity

In this section complexity will be discussed in terms of characteristics of complexity in the configuration of projects (differentiation and interdependence) and two characteristics affecting the complexity of the project management process (uncertainty and information).

2.1 Differentiation and interdependence

Projects are built systems that consist of elements and connections [1]. These constellations of elements and connections have a certain degree of complexity. Baccarini [2] describes complexity with the terms *differentiation*, i.e. the number of varied elements, and *interdependence* or connectivity, i.e. the degree of interrelatedness between these elements. The complexity of a system increases as the differentiation of its elements and the extent to which they are interdependent increases. Underground projects often have a high level of both differentiation and interdependence, due to the often complex and vulnerable working environment and the important interface with the soil.

Not only can the technical/physical domain of a project be seen as a system: its organization can as well [2, 3]. In organizational systems actors are the elements and the relations between them constitute their interrelatedness. Organizational systems can therefore also be characterized by the concepts of differentiation and interdependence.

In the remainder of this article the focus will be on project organization. A project organization consists of the constellation of actors steered by the project client to realize the underground technological artefact. The project client is the actor who is put in charge by the project owner to implement the project. The remainder of the project organization consists of engineering designers, construction contractors and consultants. In this multi-actor system there is an important distinction between *project managers* (who are usually working for the client) and *functional managers* (engineers, who are working for the other actors) [3]. The differentiation between these actors is made in terms of competence, values, interests and resources (cf. [4]). Differentiation also incurs interdependence in this system. The project management unilaterally requires the resources of the other actors. It also has decision-making authority, which the others usually do not have; the main interest of contractors, for example, is in making a profit. The focus in the remainder of this article will be on mechanisms that can be used to manage complexity in these multi-actor project organization systems. While taking effect in the operational project organization, at the early stages these mechanisms can be important considerations for the project owner when setting up a project.

2.2 Uncertainty and information

Based on Jones and Deckro [5], Williams [6] adds *uncertainty* to differentiation and interdependence as a third aspect of complexity. Galbraith [7] defines uncertainty as the gap between the information that is required and the information that is available. The challenge of realizing the technical system conditions for the information required and the organizational system should be considered on the basis of:

- The amount of information available within the project organization;
- The information processing abilities of the project management.

The pace of the growth of expertise, however, often has a hard time keeping up with the pace of the growth of complexity of underground projects. The natural response would be to involve more actors with information resources. But the uncertainty gap can also be narrowed in another way, namely by optimizing the project organization, thereby increasing the ability to process information.

3 Manageability

Given this complexity, what can project owners do to keep a project manageable? From section 2 it follows that the project organization should be able to deal with differentiation, interdependence and the processing of information. Owners commissioning a project should therefore keep the following in mind:

- How many actors can be managed?
- How potentially divergent are these actors' values, interests and resources?
- In what ways do the client, designer and contractors depend on each other for information and decision making?
- How is information in the organization processed?
- Do the information processing abilities meet the requirements for the particular technical system?

The interdependence between the project managers and the functional managers is based on the fact that functional managers generally have the most information about the system to be built and project managers have the decision-making authority In other words: project managers need the functional managers' information in order to make decisions and functional managers need the project managers' decision to apply their information in building the technical system. In this relationship there is a strong divergence of values and interests between the client (project managers) and other actors (functional managers) [8]. The main interest of most private actors is making money, whereas the interest of the client is fully connected to the project's performance benchmarks (implementation time, cost, scope, quality [9]).

Information and expertise – the main instruments to cope with uncertainty – are thus resources of strategic value that parties are not automatically willing to share. Clients depend on actors who have more understanding of the system to be built, but with not necessarily the inclination to act in alignment with the client's interests. This situation is commonly known as the principal-agent problem [10]. Manageability can, in large part, be assessed by the existence of the principal-agent problem between project managers and functional managers. Moe [11], Müller and Turner [12] and Winch [13] characterize this problem in two ways:

- *The adverse selection problem*. Bad results may occur due to information asymmetry between project managers (principals) and functional managers (agents) or between client (principal) and contractor (agent). The agents in these relationships know more about the risks involved in a project than the principals. It is therefore difficult for the principals to make good assessments. Moreover, if a project must be completed with a challenging design or a tight budget, it is likely that the winning bid is not the best option.
- *The moral hazard problem*. The project managers are only looking out for their own interests. They will do what is best for themselves and only do what is best for the owner if their interests are aligned.

4 Ownership and management

Considering that it is difficult to keep up with the information required, information processing may provide an answer to the challenges of complex underground projects. When considering the project organization as a means to achieve better manageability, the commissioner should keep the complexity of the organization's configuration – i.e. differentiation and interdependence – in mind. Particular attention will go to the interdependence of project managers (clients) and functional managers (engineers), where the management of differentiation and interdependence manifests most particularly. A few examples will show the patterns of manageability in the project organization that are relevant when considering the commissioning of a project.

4.1 Project organizations in practice

Project organizations can be broadly classed into five forms according to how they deal with differentiation and interdependence. To describe these forms, multiple actors have been simplified into three main actors that are distinguished in the project organization: the client (project manager), the engineering designer and the construction contractor (both functional managers). One party can play more than one role. In addition to this, there are often other actors involved, such as consultants and insurance companies. They are not considered part of the project organization, but may be important as a source of information. The client is the organization that is principally responsible for the performance of the project. Manageability will therefore be analyzed from the client's perspective.

First, a project owner may delegate a project to a designated project management entity, which acts as a client of both an engineering designer and a construction contractor. The engineering designer delivers a design to the client and the client procures a construction contractor to implement it. It is common that the engineering designer oversees the realization of his design by acting as a director for the construction contractor. The client remains responsible for the most important decisions on the basis of available information and the assessment of this information by the designer.

A second possibility is that the client has sufficient expertise to complete a design and to oversee its realization, and simply hires a construction contractor to build it. A third option would be to install a client who completes the design but separates the roles of designer and director by hiring an external project director. The fourth possibility is that the engineering designer and contractor roles are combined in a 'design and construct' contract. In such a case the contractor acts as its own director and the client offers only framework guidance. The fifth possibility is that all roles, similar to the fourth option but including clientship, are united in one entity. In such a case, the owner provides only a framework and leaves interpretation and completion entirely to a contractor or group of contractors.

4.2 Manageability in practice

How do different types of commissionership retain control over the project and in the meantime generate or process sufficient information and expertise to minimize the uncertainty gap, so that performance benchmarks can be met? To find out, a few practical examples of underground projects will be presented. It should be noted that it is very difficult to value the different types of project organization definitively, as every single project is unique, and complexity differs from project to project. A project that is a bit less complex and was carried out successfully has not necessarily performed better than a very complex project that has encountered certain problems. Moreover, the level to which performance benchmarks were met may influence the assessment of manageability. This section will therefore show a few *patterns* of (un)manageability. All cases were part of research by the author.

4.2.1 Separate client, designer/director and construction contractor

The most vulnerable clients are those that are fully dependent upon other actors when making decisions about the technical system. Nevertheless, they are numerous. And the more complex a project, the more likely it is that the owner will hire external designers and contractors to implement it for the simple reason that the expertise of the client falls short. Two projects will show the hazards of this situation: the Souterrain project by the Dutch city of The Hague and the Central Artery/Tunnel Project by the Commonwealth of Massachusetts in Boston (USA).

4.2.1.1 Souterrain, The Hague

The Souterrain is a three-storey underground structure in the centre of The Hague, consisting of a tram tunnel (lowest storey) and a two-deck underground car park. The municipality of The Hague was the owner of the project. It set up a project organization within its own city management department. As it was not very skilled or experienced in tunnelling, it hired a private engineering designer who would also oversee (as a project director) the implementation of its design by a private contractor.

In the procurement phase, the preferred bidder for the construction job questioned a part of the design that would seal the construction pit to be sealed during construction work. This design was relatively unproven and had, for cost reasons, been put together with limited robustness. Neither the contractor nor the insurance company wanted to accept liability for this part of the project. The engineering design firm stood by its design. This put the municipality in an awkward position, as it did not know how to weigh the comments of the contractor and the insurance company. Were they sincere or were they acting strategically, so that they would not have to accept liability for any possible risk? The municipality, which did not have the expertise to assess the technology on its own, retained the engineering designer and decided to proceed with the existing design. The contractor did not block the process, as liability could now be waived and the project was very important for them. They had won the tender with a surprisingly low bid, which was an extra reason for the client to be suspicious about requests for changes that might result in additional work.

One and a half years after the start of the implementation, the tunnel under construction was flooded as a result of a breach in its seal. In the subsequent process to find a technique to finish the work, the contractor played hardball, strongly distrusted the designer and after threatening to withdraw was allowed to finish the project with its own design and its own – expensive – technology [14].

This case shows, first of all, the difficulty that project management has in valuing input from engineers who are considered to be more skilful than they are, but it also shows another dimension of the moral hazard problem. The question is not only whether contracted experts are willing to share their information, but also whether the information from different 'agents' may be contested and therefore be difficult for managers with less engineering expertise to assess.

4.2.2.2 Central Artery/Tunnel Project, Boston

The Central Artery/Tunnel Project was a scheme to rebuild Boston's Central Artery, an elevated expressway that cut up the downtown area, repositioning it underground. The project was too extensive to be detailed here. Basically it was composed of many subprojects: a downtown tunnel, a connecting cable-stayed bridge, two consecutive tunnels under the harbour to the airport, and many additional sections. The tunnels had to be woven between many existing structures, both above ground and underground. It took about fifteen years to build all parts of the project.

Prior to the start of construction, the project owner, the Massachusetts Department of Public Works, had been downsized and was not equipped to manage the whole project by itself. Therefore, the department hired a large management consulting firm to make preliminary designs and to oversee implementation. This meant that the management consultant was supposed to oversee the contractors and designers and the Department of Public Works had to oversee the management consultant. The size and internal variety of the project's technical system resulted in 38 different section design consultants being hired and 142 construction contracts being issued [15, 16].

During the work a large number of claims and changes were filed with the project's management. In many contract areas differing site conditions were in effect and during implementation many minor design changes were made that led to changes in contracts. All the claims and changes piled up at the project management's office. Many of these necessary changes are said to have been the result of flawed designs by the management consultant [17–19]. However, as the owner did not have the expertise of the management consultant, it depended heavily on the consultant's work and could not assess on its own whether the numerous claims for changes resulted from flawed work.

In the meantime, costs grew massively during construction. This was partly caused by inflation, but also by the numerous changes, along with various other reasons. The growing cost overruns and troubled decision making did persuade the project owner to reconsider its project organization. The owner put together an Integrated Project Organization (IPO). Previously, many positions were held by employees of both the owner and the management consultant. In the IPO the most qualified person would stay and the redundant position was removed. By doing so, the two organizations were melded into each other. The owner hoped to move closer to information resources by doing so, but in practice this impeded oversight even further. It saved costs but also removed the checks and balances that were in place within the project organization. Considering that the management consultant was paid cost-plus-fee, the management consultant did have some interests that diverged from the client [17]. As a result, it has been difficult, if not impossible, to assess whether the management consultant did a good job.

This project shows another example of the difficulty a client has in managing hired engineers, particularly when processing input from many contractors, and the inability of providing oversight when it depends heavily on an actor with its own values and interests.

4.2.2 Client as designer/director

A different interpretation of commissionership is found in the Stadtbahn tunnel construction in the German city of Dortmund. In 1969 a grid of three tunnels was planned in Dortmund's inner city. The work would take place sequentially for over thirty (eventually almost forty) years and, to this end, the city decided to set up a designated Stadtbahn construction department, equipped with some eighty engineers. This department has now finished the last stretch of its work. Although specifications for this stretch were drawn up by an external engineering agency, implementation by construction contractors, direction and oversight were managed by the Stadtbahn department's in-house engineers. This department could also manage all decision making on technical issues, backed-up by a geo-technician and a specialist tunnelling engineer (from the same agency that had produced the specifications). The project managers from the Stadtbahn department and the functional managers from the construction contractors remained strictly separated throughout the implementation, in order for the Stadtbahn department to be able to provide oversight [20].

With its own expertise, the Stadtbahn department as client prevented dependency on the construction contractor's information. The most important external provider of expertise was uncoupled from the actual implementation of the project and, hence, had hardly any interests or values that diverged from the client's.

4.2.3 Designer/contractor as owner

Recent developments have added new types of commissionership to the traditional ones. Possible motives are the decentralization of expertise from government agencies to private firms in many countries, the unavailability of funding or the sharing in or exclusion from (financial) risk that can be obtained.

The Herrentunnel in the German city of Lübeck was designed, built and financed by a consortium of banks and construction firms. They could use the funding provided by the federal authorities to the City of Lübeck and will maintain and operate the tunnel for a designated concession period, after which the tunnel will be transferred to the city. In the concession period the consortium can earn back its investment by levying tolls on users. The city did not participate in the implementation and provided only a framework of conditions for a fixed link across the river Trave [21].

The benefits of this consortium are not only that the city gets a tunnel it did not have the money and expertise for, and that it does not have to cope with risks of construction, but it also reduces uncertainty since the primary decision-maker has direct input and access to information. As the interests of the banks and construction firms are similar, the construction firm no longer has an incentive to provide information strategically. This nullifies the moral hazard problem and as there is basically one actor for design, construction and maintenance, there are hardly problems of contested information.

4.3 Project organizations and the manageability problem

The inclination to include more information in project organizations to manage complex underground projects is a 'mono-actor' response to complexity. In reality, as the examples show, project organizations are multi-actor systems.

In such systems not only resources, but also values and interests may diverge. Due to principal-agent problems between project managers (client) and functional managers (designers and contractors), uncertainty may remain despite the availability of more information. The above shows that this problem can be tackled with alternative project organization set-ups. The expertise of the client is important in these set-ups. The cases show that simply hiring actors with information resources may not suffice; the ability of the client to process information in order to make decisions is important as well.

5 Conclusion

The natural response to complexity and uncertainty when commissioning and managing complex underground projects is to add more actors to provide more information. Although uncertainty is defined as the gap between the information that is required and the information that is available, more information does not always lead to less uncertainty, as the inclusion of more actors increases differentiation and interdependence and, hence, complexity. As a result more information resources do not always lead to more understanding. Information may be contested, used strategically, or difficult to process. Moreover, decision makers are usually not the actors with the most extensive information resources and the principal owners of these resources are often not the ones who make decisions. This all keeps uncertainty intact. Rather than attempting to increase information to reduce uncertainty in traditional project manager-functional manager relationships, avoidance of the principal-agent problem between those two types of managers may provide more support. When considering commissionership, owners should particularly keep in mind the way the foreseen actors in the organization will have to cooperate, how they depend on each other, and what this means for dealing with the inevitable uncertainty of these projects. A well thought through set-up may maximize manageability and thereby increase the chance of meeting project performance benchmarks. It may require project organizations that differ from the ones that are most familiar to many owners.

Acknowledgements

This publication is the result of work carried out by the Delft Centre for Sustainable Urban Areas, at Delft University of Technology.

The author wishes to thank Dr. Wijnand Veeneman for his comments and suggestions on the article.

References

[1] Dörner, D., The logic of failure; recognizing and avoiding error in complex situations, Basic Books/Perseus Books: Cambridge MA, 1997.
[2] Baccarini, D., The concept of project complexity – a review. International Journal of Project Management, **14(4)**, pp. 201-204, 1996.
[3] Cleland, D.I., King, W.R., Systems analysis and project management. Singapore: McGraw-Hill, 1983.
[4] Bruijn, H. de, Heuvelhof, E.F. ten, Management in Networks, Abingdon Routledge, 2008.
[5] Jones, R.E., Deckro, R.F., The social psychology of project management conflict. European Journal of Operational Research, **64(2)**, pp. 216-228, 1993.
[6] Williams, T.M., The need for new paradigms for complex projects. International Journal of Project Management, **17(5)**, pp. 269-273, 1999.
[7] Galbraith, J.R., Organization Design, Reading MA: Addison-Wesley, 1977

[8] Jensen, C., Johansson, S., Löfström, M., Project relationships – A model for analyzing interactional uncertainty. International Journal of Project Management, **24**, pp. 4-12, 2006.
[9] Turner, J.R., The handbook of project management; Improving the processes for achieving strategic objectives, Maidenhead: McGraw-Hill, 1993.
[10] Jensen, M.C., A theory of the firm: governance, residual claims, and organizational forms, Cambridge MA: Harvard University Press, 2000.
[11] Moe, T.M., The politics of structural choice: toward a theory of public bureaucracy. Organization theory: From Chester Barnard to the present and beyond, ed. Williamson, O.E., New York: Oxford University Press, 1995.
[12] Müller, R., Turner, J.R., The impact of principal-agent relationship and contract type on communication between project owner and manager. International Journal of Project Management, **23**, pp. 398-403, 2005.
[13] Winch, G.M., Managing Construction Projects, Oxford: Blackwell Science, 2002.
[14] Leijten, M., Bruijn, H. de, Complex decision-making in multi-actor systems. Managing Technology and Innovation, eds. Verburg, R.M., Ortt, J.R., Dicke, W.M., Abingdon: Routledge, pp. 192-206, 2005.
[15] Hughes, T.P., Rescuing Prometheus, New York: Vintage Books, 1998.
[16] Altshuler, A., Luberoff D., Megaproject: A political history of Boston's multi-billion dollar Central Artery/Third Harbor Tunnel Project, Cambridge MA: John F. Kennedy School of Government, Harvard University, 1996.
[17] US Department of Transportation Office of the Inspector-General, Impact of water leaks on the Central Artery/Tunnel Project and Remaining Risks, 2005.
[18] Bechtel/Parsons Brinckerhoff, The Big Dig: Key facts about cost, scope, schedule and management, 20 February 2003.
[19] Commonwealth of Massachusetts, Office of the Inspector General, www.mass.gov/ig/ig_bigdig_reports.htm.
[20] See for example http://stadtbahnbauamt.dortmund.de.
[21] See for example www.herrentunnel.de

Underground nuclear parks: new approach for the deployment of nuclear energy systems

C. W. Myers[1], J. M. Mahar[2], J. F. Kunze[2] & N. Z. Elkins[1]
[1]*Los Alamos National Laboratory, Los Alamos, New Mexico, USA*
[2]*Idaho State University, Pocatello, Idaho, USA*

Abstract

It is possible that hundreds to perhaps thousands of new nuclear power reactors could be deployed this century to help meet the growing global demand for electricity. Underground reactor siting is proposed as a potentially superior alternative to surface siting. Past studies and experience with underground siting proved the engineering feasibility and revealed numerous safety, security, environmental and aesthetic advantages, but in spite of these advantages the added cost associated with underground siting continues to be viewed as an impediment. Recent work on the underground nuclear park (UNP) concept, however, indicates the potential to reduce per-reactor capital and operating cost below that for conventional surface siting. In addition, under a closed fuel-cycle policy, reprocessing plant, fuel re-manufacturing facilities, fast spectrum reactor(s), and waste disposal facilities could also potentially be located underground as part of the UNP. Work to date has included underground design concepts and excavation cost estimates for UNPs in bedded salt and granite, and ideas for UNP-based energy system applications. The UNP approach has the potential to reduce many of the cost, waste management, safety, and security concerns currently associated with nuclear power.

Keywords: underground nuclear park, nuclear reactors, nuclear fuel cycle underground reactor siting, cost reductions, waste management, security, safety.

1 Introduction

Nuclear energy is increasingly recognized as an important technology to help meet the growing global demand for multi-gigawatt levels of baseload electricity

WIT Transactions on the Built Environment, Vol 102, © 2008 WIT Press
www.witpress.com, ISSN 1743-3509 (on-line)
doi:10.2495/US080071

on a reliable and sustainable basis, and to reduce air pollution and the emission of greenhouse gases. Nonetheless, divergent and often passionate views continue to be held regarding the issues of security and safety of nuclear reactors, the economics of nuclear power, and the proliferation and environmental risks associated with nuclear material and waste produced by nuclear power reactors.

The premise of this paper is that a significant expansion of nuclear power will ultimately result in response to growing recognition of the environmental and energy security advantages of nuclear power. Currently, 439 operating reactors produce 16% of the world's electricity, and an additional 34 are under construction with 365 either on-order, planned or proposed [1]. Assuming a significant global expansion, global growth could reach 1000 total deployed reactors by mid-century, according to a Massachusetts Institute of Technology study co-chaired by J. Deutch and E. J. Moniz [2], and perhaps even 8,000 to 10,000 reactors by the end of this century according to J. Ritch, Director General of the World Nuclear Association [3]. Given the possible deployment of 1000-plus reactors over the coming decades, the question arises of whether there might be a way to reduce beyond current levels the risks and costs associated with construction and operation of nuclear reactors relative to conventional, surface-sited reactors, and to increase their margins of safety, security and proliferation resistance. Underground siting is one possibility.

Underground space is an under-appreciated natural resource that could be utilized to facilitate the global expansion of nuclear power. Underground siting of several nuclear reactors and the waste management, reprocessing, and fuel re-manufacturing facilities supporting those reactors at a single location to form an underground nuclear park (UNP) could be an attractive alternative to conventional deployment and surface siting of these facilities.

This paper summarizes work to date on the UNP concept.

2 Past experience and studies of underground reactor siting

Three reactors were sited underground in 1958, 1961, and 1964 in a large granite rock mass near the Yenisey River in Central Siberia, Russia [4]. Bach [5] described small underground test and research reactors installed in the 1960s in Norway, Sweden, and Switzerland, and a small power plant in France The decades-long success of the Russian reactor operations and the European experience demonstrated the technical feasibility of operating nuclear reactors underground. Also, studies of underground reactor siting in the 1970s in the U.S.A., Canada, Japan, and Switzerland confirmed the technical feasibility and revealed many safety, security and other advantages, but concluded that underground construction could have a cost and schedule penalty [6]. Interest in nuclear power, as well as underground siting, waned in the late 1970s and 1980s in the wake of the Three Mile Island and Chernobyl accidents and growing public opposition to nuclear power. The exception appears to have been in Russia where interest in underground siting continued into the 1990s and was viewed, for example, by Dolgov [7] as being potentially economical, with advantages in operational safety and physical security.

Especially noteworthy is the fact the eminent 20th century physicists Andrei Sakharov and Edward Teller both saw nuclear energy as essential for the future of humanity and both advocated underground siting as a means to assure safety, even under extreme accident situations, and to promote greater public acceptance [8,9].

3 Underground Nuclear Park

3.1 Original UNP concept

Introduction of the underground nuclear park (UNP) concept in 2004 by Myers and Elkins [10] expanded the possible approaches to underground reactor siting. The number of reactors to be sited underground was increased from one to as many as 18, and the spent fuel storage facility and waste repository supporting those reactors was collocated underground along with the reactors in an open-fuel-cycle configuration (Figure 1). The 2004 UNP concept included high-temperature reactors and heat exchangers sited 200 meters deep in a thick, bedded-salt rock host rock. Multi-gigawatt levels of produced electricity were to be supplied to users by a high-capacity transmission system. Arguments were presented indicating that the life-cycle cost of electricity from the UNP would be less than under conventional surface siting and waste management approaches, and the level of public acceptance would be greater. Normal underground hazards such as fire and hazardous gases were recognized as needing analysis, as was a safety analysis to evaluate accident scenarios.

Studies completed in 2006 by Myers *et al* [11] indicated the UNP approach would lead to reduced per-reactor cost for construction, operations, security, and waste management relative to an equivalent number and type of conventional surface-sited reactors. Other advantages included increased margins of operational safety, security against attack, and protection against severe weather effects. Conceptual layout and preliminary excavation cost estimates were given for a hypothetical UNP with 18 reactors and their turbine/generators sited in an array of individual chambers at a depth of 100 to 300 meters in bedded salt. Collocation of waste management facilities underground with the reactors would reduce waste transportation cost, associated health and safety risks, and public concern. Environmental justice would be promoted because by collocating the reactors and waste management facilities at the same location the community that benefited economically from the construction and operation of the reactors would be the same community that accepted the waste from the reactors.

3.2 Closed-fuel-cycle UNP concept

Recently, through the Advanced Fuel Cycle Initiative and Global Nuclear Energy Partnership, the U.S.A. has begun to examine technologies for an improved closed-fuel-cycle to further reduce the risk of proliferation, minimize nuclear waste, and maximize energy recovery – as well as promote the expansion of nuclear energy in the U.S.A. and in other nations.

As a result, Mahar *et al* [12,13] developed a UNP concept for a closed-fuel-cycle based on a granitic host rock and tunnel boring machine (TBM) excavation. In a closed-fuel-cycle UNP, a reprocessing plant, fuel re-manufacturing facility and fast spectrum reactor(s) are collocated underground along with the power reactors, spent fuel storage facility and repository (Figure 1). There would be 10 – 15 power reactors, using the 1000 MWe light water reactor (LWR) as the unit. This number and capacity would produce sufficient quantities of spent fuel to justify a reprocessing plant and fuel re-manufacturing facility that, based on nominal designs, could be sized and dedicated to serving those reactors. Nuclear waste produced during reprocessing and fuel re-manufacturing would be disposed of in the UNP repository. Reactor decommissioning could be accomplished by in-place burial inside the UNP. (Although not included to date as part of the UNP concept, it is possible that underground siting of uranium enrichment facilities as part of a UNP would also be possible.)

An important implication of the closed-fuel-cycle UNP concept is that – to a first approximation – enriched uranium fuel would enter the UNP and nuclear energy would exit, and nothing else. For the reactors in the UNP, the back end of the fuel cycle would therefore be completely closed and permanently contained underground at this single location. Risks to public health and worker safety associated with conventional decommissioning and nuclear waste transport would be largely eliminated, and the cost of those activities would be a fraction of current cost.

3.3 Underground openings layout and construction

Conceptual design of the closed-fuel-cycle UNP includes a TBM-excavated, 15-meter diameter, main tunnel having a rectangular footprint with 610 meter minimum-length sides. Two 12 - 24 meter shafts provide equipment and personnel access to the tunnel. Drill and blast and/or additional TBM operations would be used to increase the height of the tunnel as needed. The tunnel is subdivided using bulkheads to create individual chambers 150 meters in length. The LWRs, steam generators, and condensers are placed in openings excavated below the tunnel invert, and the turbine/generator units are installed on the floor of the tunnel. The spent fuel pool is in a chamber adjacent to the tunnel. Independent access to the main tunnel is through adits connecting a second smaller tunnel driven parallel to the main tunnel.

4 Reduced capital and operating cost

Economic issues continue to be a constraint on the expansion of nuclear power. Although the UNP concept has inherent safety, security, and other advantages over conventional surface-siting of nuclear power plants, deployment of UNPs will probably not occur if there is a significant economic disadvantage. The economic viability of the UNP concept will be determined primarily by the life-cycle unit cost of power from a UNP relative to that from an aggregate of conventional nuclear power installations with similar capacity.

Past work identified several means by which the UNP concept might reduce capital and operating cost. Key among these are reduction of spent fuel and nuclear waste transportation cost, elimination of the containment structure, in-place decommissioning, and reduced security, maintenance, and insurance cost. Severe-weather related risks and its associated cost would also be reduced by being underground, as would earthquake risk.

The life-cycle unit-cost of power generated at a given nuclear power installation is dominated by the capital cost. Although the capital cost would be high for the first reactor installed in a UNP, the expectation is that the per-reactor cost would decrease significantly as the construction of subsequent reactors proceeded. This cost decrease would result from sharing the cost of common access shafts and tunnels among several reactors and from other factors pointed out in a University of Chicago study [14]; these include continuous construction of multiple units at a single site, use of standardized reactor designs, and maintaining a continuing on-site work force. The potential to incorporate pre-existing underground space as part of the UNP could also lower underground construction cost.

4.1 Reduced commodity cost

A recent study of nuclear power sponsored by The Keystone Center [15] noted an escalation beginning in 2003 in the cost of steel, cement and other construction materials. This could be an increasingly important issue in the cost of new surface-sited nuclear power plants because the construction of the containment structure, reactor building, nuclear fuel handling facilities, turbine-generator building, and the auxiliary buildings requires large quantities of concrete and structural steel. In contrast, creation of underground space for facilities usually involves the net removal of material (rock), not the addition of material. The rock remaining after excavation – reinforced as necessary – provides the structural support normally provided by concrete, steel, and lumber in conventional surface-sited buildings. In addition, in the underground there is no need to construct thick walls of nuclear-grade concrete for radiation shielding and containment because sufficiently-thick rock-mass walls would serve this function. Therefore, at a suitable site, less steel and concrete would be needed to create an underground nuclear park relative to the quantities required for surface-sited reactors with similar aggregate capacity. Lastly, the UNP approach offers the potential to avoid not only the use of large quantities of concrete and steel for nuclear power plant construction, but to also avoid much of the risk and economic uncertainty associated with the potential for further cost increases in these commodities.

5 UNP-based energy systems

A UNP could be the foundation for an energy system to produce baseload electricity, peaking electricity, hydrogen, oxygen, desalinated water and process heat. Off-peak electricity from the UNP power plants could be used to compress and store air underground in solution- or mechanically-mined caverns that would be released during higher demand periods for use in gas turbines to supply peaking power. This is proven technology known as compressed air energy storage (CAES), see Crotogino and Mohmeyer [16] for a description of CAES.

Electricity from the UNP could be used to produce hydrogen by electrolysis, or, alternatively, a UNP with high-temperature reactors could supply high temperature process heat to thermo-chemically dissociate water as per processes described by Uhrig [17] in his review of options for producing hydrogen using nuclear energy. The produced hydrogen and oxygen could then be stored as a compressed gas in underground caverns near the UNP from which it could be periodically withdrawn for use in fuel cells, for example, as described by Forsberg [18]. Also, it is possible that the stored oxygen could be used in gasification combined cycle (IGCC) plants enabling a pure stream of CO_2 to be produced, as described by Kunze *et al* [19], for subsequent sequestration, enhanced oil production or other uses. Waste heat from a UNP could be used for desalination of sea water, or in inland areas for desalination of water drawn from deep saline aquifers.

6 Summary and recommendation

The UNP is a new application of the earth's underground space resource. Relative to the conventional approach to site nuclear power plants at the earth's surface, the UNP approach can potentially provide lower capital and operating cost, a more equitable approach to nuclear waste management, greater levels of protection for nuclear material, and increased margins of operational safety and physical security.

Nations contemplating construction of new reactors, reprocessing facilities, or fuel re-manufacturing facilities should consider the UNP as an option, and undertake the engineering and economic analysis necessary to evaluate its merits with regard to their needs.

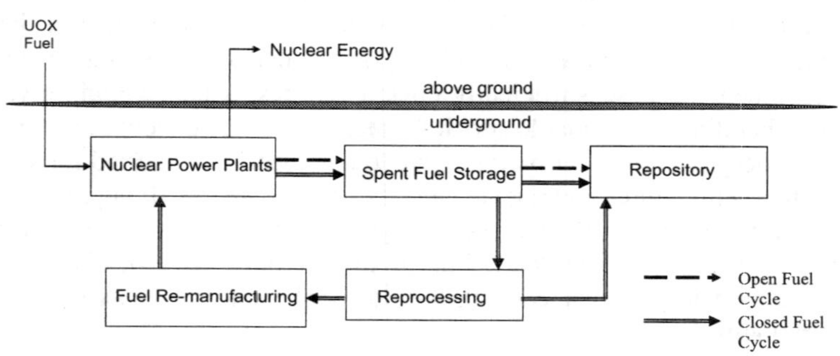

Figure 1: Underground nuclear park components and material flow relationships.

References

[1] World Nuclear Association, " World Nuclear Power Reactors 2006-2008 and Uranium Requirements, 14 January 2008," http://www.world-nuclear.org/info/reactors.html.

[2] Massachusetts Institute of Technology, "The Future of Nuclear Power," http://web.mit.edu/nuclearpower.

[3] Ritch, J., "The Necessity of Nuclear Power: A Global Human and Environmental Imperative," Speech at the World Nuclear Association-World Nuclear University, Senior Executive Dinner, London, UK, January 29, 2008.

[4] Global Security, "Krasnoyarsk-26/Zheleznogorsk Mining and Chemical Combine [MCA] N 56° 22' E 93° 41'," http://www.globalsecurity.org/wmd/world/russia/krasnoyarsk-26_nuc.

[5] Bach, P.J., "A Summary of Studies on Underground Nuclear Power Plant Siting," *Underground Space*, **2**(1), pp. 47–51, 1977.

[6] Bender, H.F., (Ed.), *Proceedings of a Symposium on Underground Siting of Nuclear Power Plants*, E. Schweizerbart'sche Verlagsbuchhandlung, Stuttgart, Germany, 412 p, 1982.

[7] Dolgov, V.N., "Inherently Safe Power-Generating Unit for an Underground Nuclear Power Plant," *Atomic Energy*, **76**(2) pp. 136–138, 1994.

[8] Sakharov, A.D., Memoirs, Alfred A. Knopf, Inc., New York, pp. 612 – 613, 1990.

[9] Teller, E., *Memoirs,* Perseus Publishing, Cambridge, Massachusetts, p. 564, 2001.

[10] Myers, W., and Elkins, N., "Siting nuclear power plants underground: Old idea, new circumstances," *Nuclear News*, **47**(3) pp. 33–38, 2004.

[11] Myers, C.W., Elkins, N.Z., Kunze, J.F., and Mahar J.M., "Potential Advantages of Underground Nuclear Parks," Proceedings of the 14th International Conference on Nuclear Engineering, 14-8913, Miami, Florida, July, 17-20, 2006.

[12] Mahar, J.M., Kunze, J.F., Myers, C.W., and Loveland, R. "Advantages of Co-Located Spent Fuel, Reprocessing, Repository and Underground Reactor Facilities", American Nuclear Society, Advanced Nuclear Fuel Cycles and Systems, Boise, Idaho, September 9 – 13, 2007.

[13] Mahar, J.M., Kunze, J.F., and Myers, C.W., "Underground Nuclear Power Parks – Power Plant Design Implications," Proceedings of the 16th International Conference on Nuclear Engineering," 16-48889, Orlando, Florida, May 11-15, 2008.

[14] The University of Chicago, "The Economic Future of Nuclear Power, A Study Conducted at the University of Chicago, August 2004," http://www.ne.doe.gov/reports/NuclIndustryStudy.pdf.

[15] The Keystone Center, "Nuclear Power Joint Fact Finding", http://www.keystone.org/spp/documents/FinalReport_NuclearFactFinding6_2007(2).pdf.

www.witpress.com, ISSN 1743-3509 (on-line)

[16] Crotogino, F., and Mohmeyer, K.U., "Huntorf CAES: More than 20 Years of Successful Operation, Spring 2001 Meeting of Solution Mining Research Institute, Orlando, Florida, April 23 – 25, 2001.
[17] Uhrig, R.E., "Producing Hydrogen Using Nuclear Energy," Proceedings International Hydrogen Energy Congress and Exhibition, Istanbul, Turkey, July 13 – 15, 2005.
[18] Forsberg, C.W., "Synergistic Benefits of a Nuclear-Renewable Hydrogen Economy," 17th Annual U.S. Hydrogen Meeting, Long Beach California, March 12 – 16, 2006.
[19] Kunze, J. F., Sandquist, G.M., and Pardo, D.M., "Is Nuclear Power Also the Key to Economically Clean Coal Gasification," Proceedings International Conference on Nuclear Engineering, Miami, Florida, July 17 – 20, 2006.

Use of a numerical model for underground stability evaluation

L. Longoni & M. Papini
Politecnico di Milano, Italy

Abstract

In recent years the tendency for re-evaluation and conversion of dismissed mines for touristic and scientific purposes (educational trails, laboratories, etc.) has been more and more diffused. To make these purposes possible in safe conditions it is necessary to evaluate the geological risk inside mines and execute proper stability studies. For the definition of geological risk, even for underground works (galleries, mines), it is not possible to prescind from the analysis of geological, geomorphological and mechanical aspects that compete for the definition of the three dimensional physical model. Only in this way will it be possible to simulate the behaviour of the rock mass and to forecast future risk sceneries, define critical thresholds and evaluate more suitable mitigation works. For a correct evaluation of the hydro geological problem using actual available means, the approach based on the physical-mathematical modelling of the problem is used more and more. This has necessitated the experimental observation of the mechanical behaviour of materials, the definition of constitutional laws, the modelling of soils and rocks systems, the formulation of holding equations, the development of analytic and numeric instruments for the solution of contour problems, the observation of the behaviour of real scale works, etc. The aim of this paper is to demonstrate the use of a numerical code for the evaluation of the stability of a part of the Piani dei Resinelli mines. The numerical model allows one to evaluate the tensional-deformational state of the rock mass. After geological, geomorphological and geotechnical analysis, we applied the 3DEC algorithm to the Piani dei Resinelli mines in order to obtain a three dimensional discontinuous model with distinct elements.

Keywords: 3DEC, hazard, mines, risk scenarios.

WIT Transactions on the Built Environment, Vol 102, © 2008 WIT Press
www.witpress.com, ISSN 1743-3509 (on-line)
doi:10.2495/US080081

1 Introduction

This study is based on a particular problem of recent years: the dismissed mines and the problem concerning the surrounding area. Sometimes these dismissed mines can generate some problems not only inside them but also in the contiguous area. For this purpose it is necessary to evaluate the stability of the mine with the definition of some risk scenarios. In the applied case shown in this paper the hydrogeological problem is a real problem because the dismissed mine analyzed is used for tourist and scientific purposes. Due to the purpose of this mine the geological hazard prediction in order to assess the safety condition is a real necessity.

The present work is part of a research with the purpose of developing a methodology for the hydrogeological analysis of the dismissed mines.

This paper focuses on a particular phase of this research: the applications of numerical modelling in underground mining. These simulations are necessary when some hydrogeological problems are possible but are an imperative condition when people, not usual in underground spaces, should enter in the mine for tourist purposes.

To assess the limitations and advantages of numerical modelling prediction, different simulations have been considered.

Numerical modelling has been used to investigate a variety of problems in underground mining and tunnelling: subsidence induced by longwall coal mining; stress generated when an open stope is filled with cemented backfill and the stability of exposures created during subsequent mining of adjacent stopes; the interaction of two tunnels; the effects of under-mining a pre-existing tunnel and shaft [2]. In this case some simulations for evaluating some possible scenarios were also made to assess the mine entrance for tourists. The real problem is that the abandoned mines are a real temptation for tourists but these spaces are very dangerous and only mining engineers, ready for the worst, have any business going into one of these mines. This paper shows the numerical models used to evaluate the real dangers in a mine, used as a prototypal case situated in Lecco.

2 Geological and geomorphologic context

Numerical modelling can assist geologists and engineers in underground excavation problems. It is important to understand the available data before starting with simulations because for high-quality simulations extensive geological and geotechnical data are necessary. Evaluation of underground instability mechanisms needs a good understanding of the geological setting, material behaviour, and physical mechanisms, as well as the use of adequate, flexible computational models to make the predictions.

2.1 The case study

With so many abandoned mines in Lombardia, it is natural to want to see what a mine looks like. Unfortunately, abandoned mines are very dangerous. Every year people die exploring abandoned mines. Fortunately tourist mines and the mining museum offer safe tours for people who are not experts in underground spaces. For safe tours through the world of mining it is necessary to analyze every risk scenario. This research deals with the evaluation of hydrogeological risk with the use of numerical models.

For the purpose of this paper the Piani dei Resinelli mines are analyzed.

These mines, shown in fig. 1, are used for tourism. Every year a huge amount of people, especially students, visit the mines as tourists.

Figure 1: Piani dei Resinelli mines.

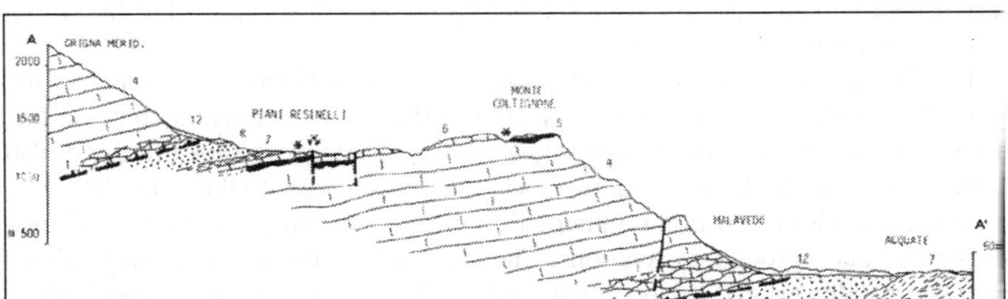

Figure 2: Geological cross-section from North to South. This section crosses the mine area (Rodeghiero et al. [6]).

2.2 Area geological classification

Some authors (Rodeghiero et al. [6]) have studied this area from the geological point of view. In this paragraph a synthesis of these previous studies are presented. Figure 2 shows the surrounding area of Piani dei Resinelli.

In the studied area sedimentary triassic rocks outcrop: the "Calcare Metallifero Lombardo" and the "Calcare di Esino". The area is affected by some folds and faults. The Calcare di Esino is etheropic with Calcare di Perledo-Varenna. It consists of grey limestones and dolomites. These sedimentary rocks are affected by close fissuration, tending to divide into blocks, and by karst phenomena. Some joints and faults subdivide the bedrock in more or less dislocated blocks. For the mine it is important to analyze the transition facies among the Esino limestone and the Metallifero Bergamasco limestone.

3 Numerical model

Up to now numerical models have often been used for the scope of civil engineering.

Although based on scientific "first principles", complex numerical models inevitably require simplifications, judgment calls, and correction factors. These subjective measures may be entirely acceptable so long as the model matches the available data, acceptable because the model is not intended to be internally consistent with all the laws of physics, but rather to serve as an expedient means to anticipate behaviour of the system in the future. This paper shows the importance of the 3D model reconstruction for simulation. It is really important to analyze all data for a real or similar representation of the analyzed model.

It is important to define two different simulation typologies:

-if a great amount of geological and geomorphologic data are available it is possible to reconstruct a 3D model similar to the real case. In this case the simulations can predict deformations and stability;

-if not, the numerical model can be used to perform parametric studies. In this case it could be possible to have a series of possible risk scenarios but first of all it is necessary to define the key parameters.

For the case study we have enough data for a good 3D model reconstruction. With this model the simulations can predict the future behaviour, but for this study it is also important to understand the key parameters. For these we also made an analysis of the different results with some changes to the input data.

There are a lot of different numerical models for the stability analysis of rock slopes and underground excavations. The analysis technique chosen depends on both site conditions and the potential mode of failure, with careful consideration being given to the varying strengths, weakness and limitations inherent in each methodology [3].

For the hydrogeological problems of the Resinelli mines a 3D numerical model was used.

As we will later demonstrate, the use of the 3D model can significantly increase the future behaviour forecast.

Due to hydrogeological problems inside the Resinelli mine the 3DEC model was chosen.

3.1 Program 3DEC

This program is provided by Itasca Consulting Group, Inc. It is a 3D numerical program based on the distinct element method for discontinuum modelling. 3DEC simulates jointed rock mass subjected to either static or dynamic loading. The discontinuous medium is represented as an assemblage of discrete blocks. The discontinuities are treated as boundary conditions between blocks; large displacements along discontinuities and rotations of blocks are allowed. Individual blocks behave as either rigid or deformable material. Deformable blocks are subdivided into a mesh of finite difference elements, and each element responds according to a prescribed linear or nonlinear stress-strain law. The relative motion of discontinuities is also governed by linear or nonlinear force-displacement relations for movement in both the normal and shear directions 3DEC has several built-in material behaviour models, for both the intact blocks and the discontinuities, which permit the simulation of responses representative of discontinuous geologic, or similar materials. 3DEC is based on a "Lagrangian" calculation scheme that is well-suited to modelling the large movements and deformations of a blocky system [5].

4 Modelling applications

This part of the paper models the underground space in Piani dei Resinelli.

The simulations have two different purposes. The first is to simulate the stability of the tunnel in order to define the risk scenarios and the other type of simulations were made to analyze the most critical parameters and the influence on stability.

Before analyzing the simulation results it is important to develop a good 3D model with all the features and boundary conditions.

4.1 The 3D model of the Piani dei Resinelli mine

By an analysis of geological data, a field survey and geomorphological examination, it is possible to build up a 3D physical model of the tunnel.

The most important phases for the 3D model reconstruction are:

Phase 1: Model generation
In this phase it is necessary to define the model geometry. It must represent the physical problem to a sufficient extent to capture the dominant mechanisms related to the geological structure in the region of interest. It is important to define the detail for the geological structure. Figure 3 shows the mine plant and the consequent reconstruction of the geometry of the tunnels.

Then, for model creation, it is necessary to generate joints. Some data were analyzed and then the most important set of discontinuities were chosen for the model reconstruction.

In this phase material properties were also defined for both the intact blocks and the joints set. Tables 1 and 2 show these material properties.

Phase 2:Deformable or rigid blocks

This is an important aspect of this program. It is necessary to decide which kind of block simulates in the best way the real behaviour of the mine. In this case the rigid blocks option was chosen.

Phase 3:Boundary and initial conditions

For boundary and initial conditions the situation of the mine was analyzed. It is important to define the vertical stress on the underground roof. It could change in the same tunnel so for this case all the different stresses were reported in each track of the tunnel. For the case study analyzed in this paper, the track of the mine is located at different depths:

-at the mine entrance it is 0 m in depth;

-at the end of the track it is 15 m in depth.

4.2 Simulations

Knowing the tunnel model it is possible to start with the simulations. Before tunnel generation some simulations were made to find the initial equilibrium of the block with the joints sets. With the initial equilibrium state the tunnel generation can be performed. Through a simplification of the real situation the tunnels were generated into the models.

Figure 4 shows the unbalanced force after the tunnels excavation. As it is possible to see, after 4000 cycles the model reaches equilibrium.

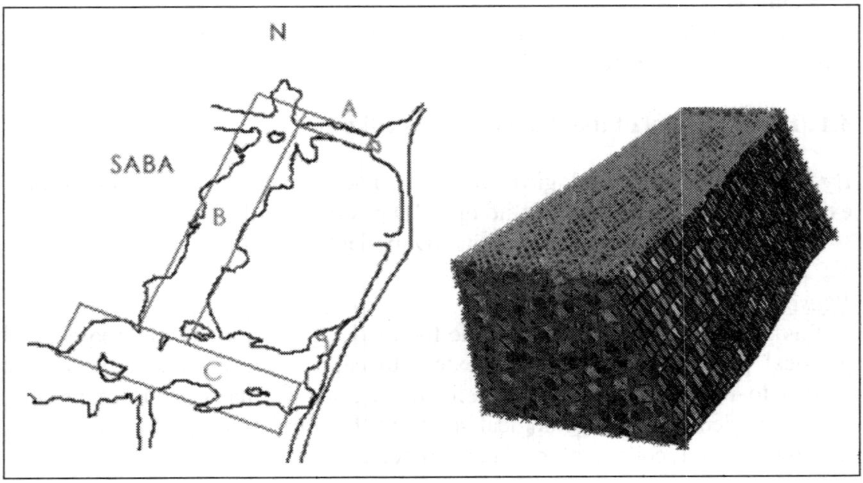

Figure 3: Model geometry.

Table 1: Block properties.

BLOCK PROPERTIES				
mass density	→	ρ	2400	N/mc
friction angle	→	φ	33	°
cohesion	→	c	3,5	MPa
shear modulus	→	G	12000	MPa
Young's modulus	→	E	30000	MPa
Poisson's ratio	→	v	0,25	

Table 2: Joint properties.

JOINTS					
Joints sets	**joint normal stiffness kn**	**joint shear stiffness ks**	**linear persistence**	**cohesion**	**friction angle**
	Pa	***Pa***	***%***	***Pa***	**°**
SO	$5\ 10^9$	$5\ 10^9$	85	$0{,}525\ 10^6$	30
K1	$5\ 10^9$	$5\ 10^9$	50	$1{,}75\ 10^6$	30
K2	$5\ 10^9$	$5\ 10^9$	30	$2{,}45\ 10^6$	30
K3	$5\ 10^9$	$5\ 10^9$	50	$1{,}75\ 10^6$	30
F1	$1\ 10^9$	$1\ 10^9$	100	0	30
F2	$5\ 10^9$	$5\ 10^9$	60	$1{,}4\ 10^6$	30

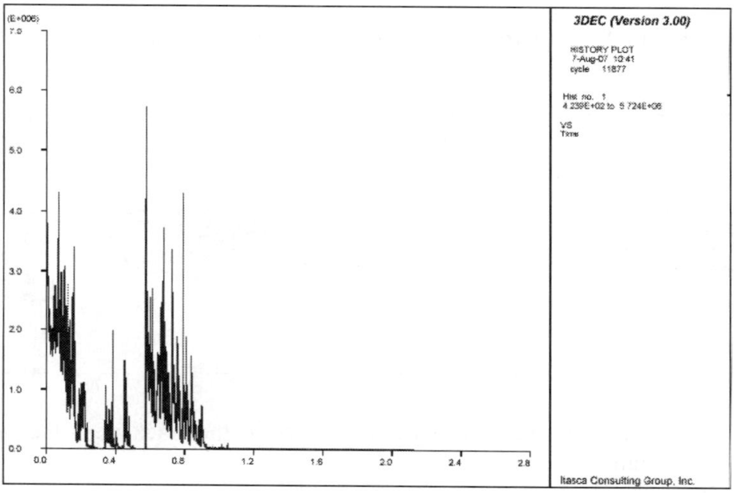

Figure 4: Unbalanced force.

The velocity and displacement history plot graphics were examined. The value is about 10^{-6} m for displacements and 10^{-6} m/s for velocity. During the simulations different parts of mine were analyzed. As reported in figure 3 the tunnels are called: tunnel A, tunnel B and tunnel C.

Cross section plots of each tunnel were generated. Figure 5 shows one of these. There is a vertical distribution of displacements but this is not dangerous for the limited value.

Instead it is important to analyze the behaviour in the track where a fault is present. Figure 6 shows the displacements and the velocity vectors of this track. It is really dangerous to have the same vertical direction of the velocity and displacement vectors, but also in this case, the situation is not critical due to the limited value of vectors.

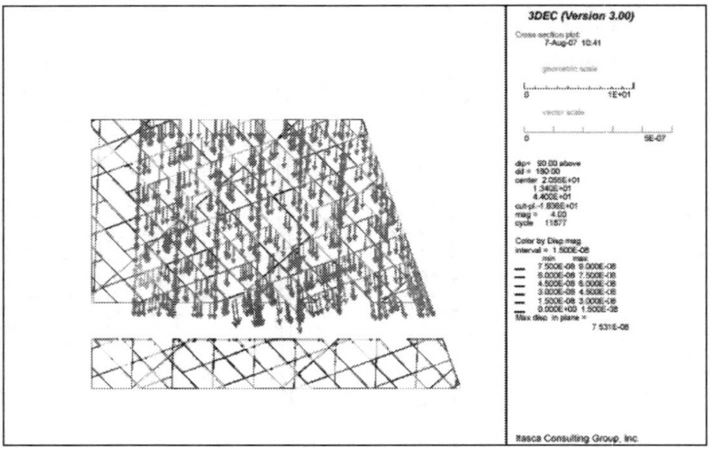

Figure 5: Cross section tunnel A: displacement vectors.

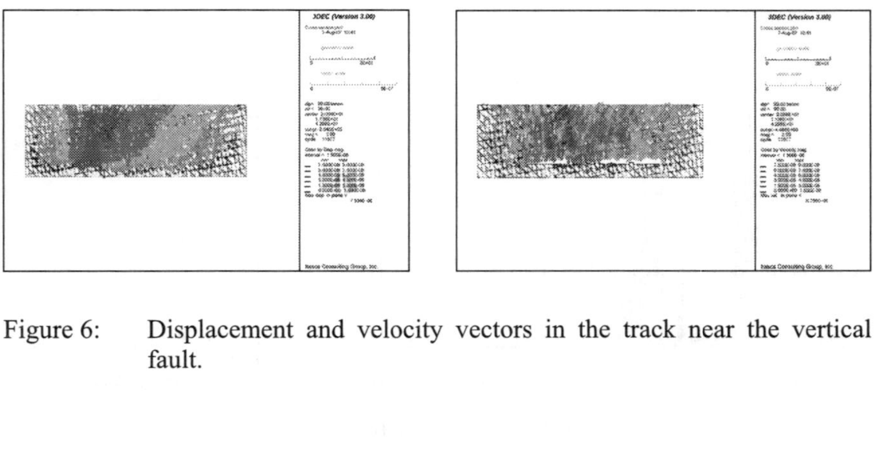

Figure 6: Displacement and velocity vectors in the track near the vertical fault.

4.2.1 Persistence analysis

During simulations it is important to define the most critical parameters not only for the risk scenarios generation but also for a best definition of site investigation and material testing.

One of the most difficult to define is persistence. Through lots of different simulations a judgment of this parameter was made.

A comparison of the different simulations show that the persistence is not a critical parameter for the failure mechanisms of the Piani dei Resinelli mines. The values of displacement and velocity vectors are the same as the previous simulations.

The analysis of this parameter was made only because the persistence is very hard to investigate.

4.2.2 Cohesion analysis

To identify the collapse mechanism the cohesion was changed. This is one of the most critical parameters for stability problems, especially for the types of blocks formed in tunnels. A lot of different simulations were made with a continuum reduction of cohesion.

Table 3 shows the results of the simulations.

Table 3: Cohesion analysis.

Cohesion reduction	displacement	velocity
10%	10-2 m	10-7 m/s
20%	10-2 m	10-4 m/s
30%	10-3 m	10-5 m/s
40%	10-3 m	10-5 m/s
50%	10-2 m	10-1 m/s

The value of the displacements and velocity is dangerous.

In the last simulation the most critical parameters were joined in order to define the failure mechanism. With an average persistence, common at all discontinuities, and a reduction of cohesion at 1.75 MPa value, some movements were registered. The collapse mechanisms defined are:
-the sliding blocks due to the presence of a k2 joint set;
-movements around the vertical fault.

5 Conclusion

In this paper we have shown the improvement in risk evaluation using a numerical model. 3DEC is a great advance in numerical approaches. It allows one to simulate the underground space in three dimensions. One of the most critical parts is the reconstruction of the 3D physical model. In this research standard methodologies must be defined in order to evaluate the hydrogeological risk in underground mines used for tourist purposes.

The real goal is to define specific risk scenarios with the use of numerical models. In this case the simulations show that the mine is stable. But if some changes in cohesion and persistence are possible the failure mechanisms defined by simulations are sliding and movements near fault. Through proceeding interactively it is possible to define detailed scenarios and so guide geotechnical design work in underground mining and construction.

Another important item is to evaluate the simulation results through warning signs observed in the mine. For this purpose a lot of monitoring systems were located in the most critical parts of the three tunnels analyzed. These instruments allow one to measure the displacements (crack meters) but also accelerations (some MEMS were located on the critical blocks to measure acoustic emission, which can then forecast new movements).

References

[1] Asche, H., The prediction of ground settlements and the design of support systems in shallow tunnels in weak rock. Breaking New ground, Sydney 1996, pp161–173.

[2] Coulthard M.A., Applications of numerical modelling in underground mining and construction, Geotechnical and Geological Engineering 17: 373–385- 1999.

[3] Doug Stead et al, Advanced numerical techniques in rock slope stability analysis – applications and limitations, conference: Landslides – causes, impacts and countermeasures. 17–21 June 2001 Davos.

[4] Hoek E., and Bray J., Rock slope engineering, revised third edition, London, 358, 1981.

[5] Itasca Consulting Group, Inc, 3DEC Manual: User's Guide. 2003.

[6] Rodeghiero F., Jadoul F., Vailati G. e Venerandi I. (1986) – Dati preliminari sulle mineralizzazioni a Pb-Zn dell'area tra Mandello e Ballabio (Lombardia Centrale). Memorie della Società Geologica Italiana, 32: 133-150.

Tunnel face stability as a function of the purchase length

J. Trckova, P. P. Procházka & S. Peskova
Institute of Structure and Mechanics of Rock,
Czech Academy of Sciences and Czech Technical University in Prague,
Civil Engineering, Prague, Czech Republic

Abstract

In geomechanical engineering the stability of the tunnel face appears to be one of the most decisive items in the list of assessments needed for verification of bearing capacity of the system tunnel – surrounding rock. Studies following from on site measurements are very expensive and depend strictly on the nature of material being involved in the study. Experiments conducted on scale models in stands which are filled by physically equivalent materials are more promising and complex, as instrumentation of smaller samples is richer and more flexible than in the case of treatment on real site. Moreover, numerical modeling, which requires a particular description of material behavior of the structures coming into the computational models, can be derived from physical models in an easier way than from real tunnel behavior. A numerical approach is proposed in such a way that internal parameters of a mathematical physically nonlinear model are evaluated using partial results of experimental scale models from equivalent materials. Although basically strongly nonlinear problems are solved, solution of linear algebraic equations is the final step of the approach for identification of values of internal parameters characterizing material properties in mathematical simulation of reality. Tunnel face stability in dependence of a length of work-out space is solved in this paper using coupled modeling, as an example of application of the procedure envisaged.
Keywords: fiber reinforced concrete lining, tunnels, coupled modeling, eigenparameters, effect of tunnel lining stiffness.

WIT Transactions on the Built Environment, Vol 102, © 2008 WIT Press
www.witpress.com, ISSN 1743-3509 (on-line)
doi:10.2495/US080091

1 Introduction

In assessment of underground structures a combination of experiments and mathematical treatments are of great interest for designers and researches. Such a coupling enables one to improve information needed in numerical modeling on mechanical behavior of structural elements being employed in the problem at relatively low cost. The following approaches can be distinguished in practice:
- Convergence analysis, comparing results from experiments and pilot numerical analysis and successively adjusting material parameters in such a way that the numerical results are in reasonable agreement with the experimental.
- Back analysis, or coupled modeling is defined as a process, in which a qualitative and quantitative measure of agreement with experimental results is ensured in mathematical model and suggests approaches, by virtue of which internal parameters of different kind are adjusted to be in compliance with experiments as close as possible. In most cases fitting of physical laws (generalized Hooke's law, creep, relaxation, aging, etc.) is sought, but sometimes new geometry arrangement is required.

Previously Cividini, et al., [1] suggested a successful approach leading to a comparative study of the rock and tunnel lining behavior and the reality in terms of internal parameters (material properties). The paper presents a discussion on some of the aspects of parameter "characterization" problems (or back analyses) in the field of geomechanics.

In 1992 Dvorak established Transformation Field Analysis, [2], which expressed nonlinear problems in a hull of linear effects and effects of eigenparameters. This method appeared a powerful tool for solving optimal prestress of a thick-walled composite cylinder consisting of many different cylindrically orthotropic layers being loaded by uniform, axisymmetric tractions and by piecewise uniform eigenstrains in the layers, [3]. The first attempts have been done in [4] to involve the eigenparameters in the coupled modeling. The starting point for this approach was elastic state and effect of eigenparameters was expressed by combination of products of influence matrices and eigenparameters. In this paper sophisticated experimental approach leading to an observation of effect of work-out space to displacements of the tunnel face is presented. The displacements can then be used in the coupled modeling.

2 Numerical modeling

As said in Introduction, the idea of the back analysis used here starts with similar assumptions and approaches as in Transformation Field Analysis (TFA). This idea is basically very simple. Consider a generalized Hooke's law, i.e. let us relate overall stresses σ with overall strains ε and eigenstrains μ, or eigenstresses λ, which may be realized as generalization of the influence of temperature: $\sigma = \boldsymbol{L}(\varepsilon - \mu) = \boldsymbol{L}\varepsilon + \lambda$, where $\boldsymbol{L}$ is the purely elastic material stiffness matrix. This generalization differs from description of the temperature in such a way that the temperature appears in Hooke's law as the trace in the strain tensor while eigenstrains or eigenstresses are full-value tensors of the second order.

In order to formulate the general procedure for the TFA, it may be done in terms of many modern numerical methods. First, let us consider that the body (part of a structure, element, and system of more elements, composite, rock, soil) behaves linearly, i.e. Hooke's linear law is valid in the entire body. When the problem is correctly posed, the displacement vector, strain and stress tensors can be obtained from the Navier equations, kinematical equations, and linear Hooke's law.

In the second step we select points, where the measured values are available, either from experiments in laboratory, or from "in situ" measurements. We also select points, or regions (subdomains) from the body under study, and apply there successively unit eigenparameter impulses (either eigenstresses or eigenstrains) to get an influence tensors (matrices). In order to precise this statement, denote A_i, $i = 1,...,n$, either the points or regions where the eigenparameters will be applied. Let, moreover, the set of points where the measured values are known, be B_j, $j = 1,...,m$. Then the real stress at B_j is a linear hull of stress $\boldsymbol{\sigma}^{ext}$ at B_j due to external loading and eigenstrains $\boldsymbol{\mu}$ (or eigenstresses $\boldsymbol{\lambda}$) at A_i (similar relations are valid for overall strain field ε) leads us to relations as

$$\boldsymbol{\sigma} = \boldsymbol{\sigma}^{ext} + \boldsymbol{P}^{\sigma} \boldsymbol{\mu}, \quad \text{or} \quad \boldsymbol{\sigma} = \boldsymbol{\sigma}^{ext} + \boldsymbol{R}^{\sigma} \boldsymbol{\lambda}, \tag{1}$$

$$\boldsymbol{\varepsilon} = \boldsymbol{\varepsilon}^{ext} + \boldsymbol{P}^{\varepsilon} \boldsymbol{\mu}, \quad \text{or} \quad \boldsymbol{\varepsilon} = \boldsymbol{\varepsilon}^{ext} + \boldsymbol{R}^{\varepsilon} \boldsymbol{\lambda}, \tag{2}$$

where the influence tensors, $\boldsymbol{P}$ and $\boldsymbol{R}$, are obtained from the unit impulses of eigenparameters introduced on selected regions A_i, $i = 1,...,n$. Note that in 3D the eigenparameter tensor is symmetric, so that six components are available as the design parameters in one region. The components of influence matrices are created from responses of stresses or strains in elastic medium again. The dimensions of $\boldsymbol{\sigma}$, $\boldsymbol{\sigma}^{ext}$, $\boldsymbol{\mu}$, and $\boldsymbol{\lambda}$ are $m \times 6$ (because of symmetric stress and strain tensors) and the dimensions of $\mathbf{P}$ and $\mathbf{R}$ are $m \times 6 \times n$.

Note that it holds: $\boldsymbol{\lambda} = -\boldsymbol{L}\,\boldsymbol{\mu}$. The free, or design parameters, which should be selected in such a way that the measured and calculated values are mutually as close as possible, can be determined from many approached. One of them is suggested in the next text, starts with formulation of an optimization problem, and can be considered a back analysis procedure.

Without any details we can assert that similar relations as that of (1) and (2) can be written for displacements:

$$(u_i)^k = (u^{ext}{}_i)^k + \sum_{j=1}^{6} \sum_{l=1}^{m} (R^u{}_{ij})^{kl} (\lambda_j)^l, \quad i = 1,...,6, \; k = 1,...,m \tag{3}$$

On the other hand measured stresses $(\sigma_i^{meas})^k$, or measured displacements $(u^{meas}{}_i)^k$ are available in a discrete set of points (namely the points B_i). A natural requirement is that the values of measured and computed values be as close as possible. This leads us to the optimization of an "error functional"

$$I[(\lambda_j)^l] = \sum_{i=1}^{6} \sum_{k=1}^{m} [(\sigma_i)^k - (\sigma_i^{meas})^k]^2 \rightarrow \text{minimum}, \tag{4}$$

or

$$I[(\lambda_j)^l] = \sum_{i=1}^{6} \sum_{k=1}^{m} [(u_i)^k - (u_i^{\text{meas}})^k]^2 \quad \rightarrow \text{ minimum} \tag{5}$$

Differentiating I by $(\lambda_\alpha)^\beta$ yields

$$\sum_{l=1}^{n} (A_{\alpha j})^{\beta l} (\lambda_j)^l = Y_\alpha^\beta, \quad \alpha = 1,...,6, \ \beta = 1,...,\text{m}, \tag{6}$$

where

$$(A_{\alpha j})^{\beta l} = \sum_{i=1}^{6} \sum_{k=1}^{m} (R_{ij})^{kl} (R_{i\alpha})^{k\beta},$$

$$Y_\alpha^\beta = -\sum_{i=1}^{6} \sum_{k=1}^{m} (\sigma_i)^k - (u^{\text{meas}}{}_i)^k + \sum_{j=1}^{6} \sum_{l=1}^{m} (R_{ij})^{kl} (\lambda_j)^l (R_{i\alpha})^{k\beta}$$

3 Physical modeling

Physical modeling is important for study of effects taking place in rock material in connection with construction of underground structures. The modeling allows us to investigate mechanisms of geotechnical phenomena, predicts stress changes and their demonstration during various progresses of underground construction and also during simulation of operating conditions.

Basic rules of the experimental modeling and formulation of the boundary conditions for modeling comes out from the principles of geometrical and physical similarity which is inferred for a consideration of dimensional analysis, [5]. Such a similarity modeling applied to slope stability is published in [6], for example. To simplify the solved problem constitutive relevant quantities v_1, $v_2,...,v_n$ can be selected, which posses exercise decisive influence to process taking place in the rock material. We assume that influence of the other quantities is lesser. Then physical equation involving function of relevant quantities of various dimensions

$$F(v_1, v_2, \ldots, v_n) = 0, \tag{7}$$

describes in simplification, given by selection of these quantities, behaviour of the rock material. According to the Buckingham theorem this dimensional equation for relation between reality and model can be reduced to the problem of finding $k < n$ relevant non-dimensional parameters π_i . They are functions of v_i , fulfil the above equation (7) and are numerically identical for model and reality. By implementation of non-dimensional parameters π_i , for which from requirement of dimensional homogeneity follow

$$\pi_i = v_1^{x_{1i}} \ v_2^{x_{2i}} \ \ldots \ v_n^{x_{ni}}, \ (i = 1, 2, \ldots, k)$$

non-dimensional physical equation is obtained

$$F'(\pi_1, \pi_2, \ldots, \pi_k) = 0,$$

in which arguments π are dimension independent. Non-dimensional parameters π_i correspond to basic central processing units ***L*** (length), ***T*** (time), ***M*** (mass).

Physical model has to obey geometrical similarity; in reality it is a proportionality of dimensions and angles between model and modeled object in the whole range of the model.

In the modeled geotechnical problems the ratio of length dimensions of the model and reality plays very important role (and is given from the intended geometry of the scale model). If $1/\alpha_l$ is the length scale, i.e. ratio of lengths in the model and reality and a is the ratio of bulk densities

$$a = \rho_{\text{model}}/\rho_{\text{reality}},$$

then we can define ratios for the following quantities

forces $P_{\text{model}} = a.\,(1/\alpha_l)^3\,.\,P_{\text{reality}}$

stresses $\sigma_{\text{model}} = a.\,(1/\alpha_l)\,.\,\sigma_{\text{reality}}$

deformations $\varepsilon_{\text{model}} = a.\,(1/\alpha_l)\,.\,\varepsilon_{\text{reality}}.$

Time as such asserts oneself during derivation of non-dimensional parameters determined from a system of equations in relevant variables and corresponding basic units. To determine time scale ratio of the time in reality and in the model empirical estimation is applied (in terms of time needed for stress redistribution in the model body caused by pressure changes in model in comparison to time when the same process took place in reality). This time scale is applied for assessment of the rest quantities depending on time. Generally time scale can be determined as

$$\alpha_t = (v_{\text{model}}/v_{\text{reality}}).\,\alpha_l,$$

where

α_t – time scale
α_l – length scale
v – velocity of deformation.

To simulate the most perfect processes taking place in the rock material, rock environment is replaced in the model by equivalent materials their determinate physical and mechanical properties according model laws and scale of model agree with rock properties and respect the character of failures simulating those in rock material. The models are constructed from mixture of various, mostly easy available materials (e.g. sand, bentonite, ballotine, gypsum, mica - vermiculite, composite mortar, cellular concrete and water).

The models are constructed in stands of various dimensions in dependence of solving problem and length scale of the model.

4 Example

The experiments are focused on physical models on a scale 1 : 100, in a model stand. In the models rock material is substituted by physically equivalent materials, which consist of various mixtures of sand, bentonite and fat (65% + 29% + 6%), and moisture. Their properties are determined by standard tests:

volume mass ρ	1.45 g/cm^3
compressive strength σ_t	0.027 MPa
strength in simple tension σ_c	0.006 MPa

cohesion c_p 0.014 MPa
angle of internal friction ϕ_p 34°50′
Young´s modulus E_{def} 0.12 MPa
Poisson´s coefficient ν 0.28

The applied equivalent material is very plastic with a high rate of permanent deformation and on a scale of the model it produced rock of a mudstone type. In order to measure movements of the tunnel lining and surrounding rock needles with small discs at one end facing to the tunnel heading, which serve for non-penetrating of needles to the rock, are installed to the modeled tunnel, see Fig. 1.

Figure 1: View of the model stand with dimension 250 x 250 x 250 mm equipped by displacement measuring system (needles).

A lined tunnel of cylindrical shape with an internal diameter of 76 mm was modeled parallel to the bottom of stand. To determine stability of the tunnel face in dependence of the length of work-out space, a part of the tunnel lining in the model test was formed from five 20 mm long rings. Among them 2 mm spacing were retained (Fig. 2). These five rings were equipped with locking mechanism, which could be released by using strings. A rubber band, fixed on the external ring circumference, closed the gap in the ring (Fig. 3). During the model experiment, the rings were successively released by a special technology described in the sequel to simulate advancement of the tunnel face. For example, the ring near the tunnel face was released to model starting instant of the excavation of tunnel opening.

Figure 2: Equivalent of the tunnel lining with inner diameter of 76 mm created by hardened paper used in the model experiment.

www.witpress.com, ISSN 1743-3509 (on-line)

As mentioned above, special equipment (based on a system of needles) is developed, which enables one to determine the values of the displacements with the accuracy of 0.05 mm (Fig. 4). Before the experiment tests the needles are lent against the tunnel face. Using a thedolite that is placed about 2 m from the model stand transversaly to the needles, the changes of the position of needles – their displacements due to tunnel face movement – are determined with respect to the initial state of the needles.

Figure 3: One ring used for the tunnel lining forming.

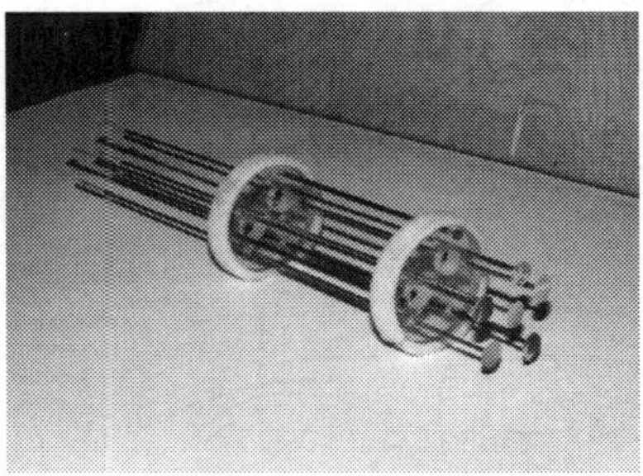

Figure 4: Aluminum needles for tunnel face deformation measurement equipped with square discs.

In Fig. 5 (a) to (e) the contour lines of displacements among the points the needles which are affected by the tunnel face movements, which are measured after successive release of the first to fifth rings of the simulated tunnel lining are shown. The length of work-out space ranges from 22 mm to 110 mm. After releasing the fifth ring the time-dependent behavior of model material is studied. Displacements measured after 50 minutes, 16 hours, 24 hours, 40 hours, 47 hours and 5 days after releasing the fifth ring are shown in Fig. 6 (g) to (k).

5 Conclusions

In this paper the procedure for calculating a laminated arch is suggested in cylindrical coordinates $0r\theta z$, starting with the assumption of generalized plain strain. Before introducing this assumption, the displacements are developed into

Fourier's series in the time and hoop coordinates. Employing kinematical equations in cylindrical coordinates, and Hooke's law the stresses are derived in the split formulation, from which the radial and axial coordinates on one hand side and the time and hoop coordinates on the other side are separated. After this variational formulation follows and finite element-like procedure is employed in the coordinate system $0rz$. In radial direction linear approximation of displacements is supposed and in the sense of the generalized plane strain also linear distribution of displacements in axial direction is introduced. Simply supported segment is considered in our case, but more general supports can be involved using given moments at the end points, the clamped edge can be simulated, for example.

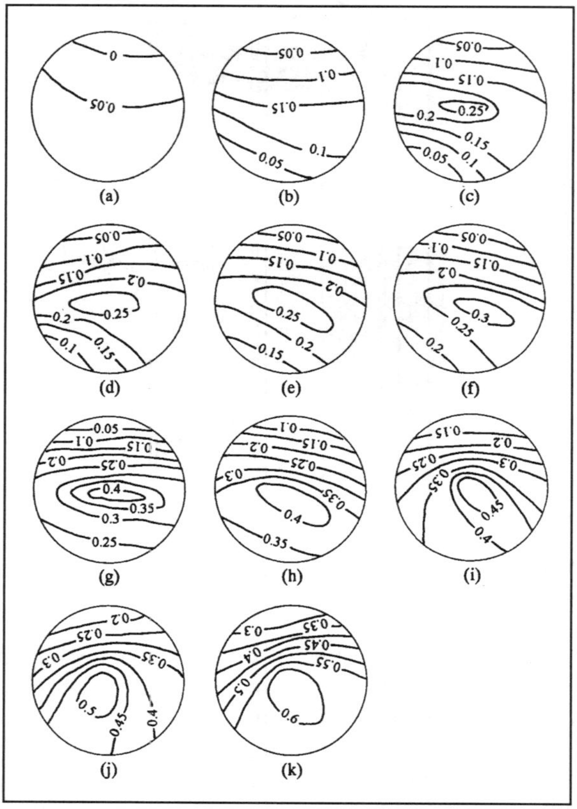

Figure 5: Contour lines of displacements in millimeters measured at the tunnel face during model experiment after simulation of the extended length of the work-out space.

As an example of application of the above described approach a dumping layer for dissipation of energy after application of explosive load is considered in various laminates. Two first natural (eigen) frequencies are observed dependent on the positioned in the structure of the arch. It appears that the most promising case is that, which is defined by positioning the dumper to the outer boundary.

Acknowledgements

This paper was prepared under financial support of GAČR, project No. 103/08/1197. The second and the third authors were partly sponsored by CIDEAS.

References

[1] Cividini, A., Jurina, L., Gioda, G., 1981. Some aspects of 'characterization' problems in geomechanics. International Journal of Rock Mechanics and Mining Science & Geomechanics Abstracts. 18, 6, 487–503.
[2] Dvorak, G.J., 1992. Transformation field analysis of inelastic composite materials. Proc. R. Soc. London A 437, 311–327.
[3] Dvorak, G.J., Procházka, P. 1996. Thick-walled composite cylinders with optimal fiber prestress. Composites, Part B, 27B, 643–649.
[4] Procházka P., Trčková J., 2000. Coupled modelling of concrete tunnel lining. In: Our World in Concrete and Structures, Singapore, 125-132.
[5] Kožešník, J. 1983., Theory of similarity and modelling. Academia, Prague.
[6] Koudelka, P., Procházka, P., 2001. Apriori Integration Method – Analysis, similarity and optimization of slopes, Academia Prague.

3-dimensional mesh generation using the Delaunay method

R. Hoshiko & M. Kawahara
Department of Civil Engineering, Chuo University, Japan

Abstract

The purpose of this research is automatic generating 3-dimensional finite element mesh using the Delaunay method. In the present age, numerical analysis is used in various problems. The finite element method is one analytical approach using the finite element mesh. But, there are problems in that large scale finite element mesh generation takes a lot of time and cost. As for the large scale finite element mesh, data are enormous, and it is almost impossible for generation only by manual labor. In addition, the mesh generators prevail, but those do not always generate meshes in response to the demand of the user. It is necessary to generate mesh for the user's purpose. The Delaunay method is a technique to divide the domain into tetrahedra by using a set of nodes that are distributed arbitrarily. This technique can be applied to generate finite element meshes of complicated configuration, that is to say, practical models. In this research, generating a Suemune tunnel and mountain-shaped analysis model using the Delaunay method is presented. But, the Delaunay method cannot dispose the concave faces and interior boundaries. When generating mesh without device, inadequate elements are generated on concave faces and interior boundaries. The mesh of the concave faces and interior boundaries cannot be generated only by the Delaunay method. These problems can be solved by Virtual nodes. The Virtual nodes represent the nodes that are generated to remove inadequate elements.
Keywords: Delaunay method, Super tetrahedron, Virtual nodes.

1 Introduction

In recent years, the performance of the electric computer has drastically improved and the technology of numerical analysis is highly progressed. Therefore it is possible to carry out large scale numerical simulation. For example, control water

www.witpress.com, ISSN 1743-3509 (on-line)
doi:10.2495/US080101

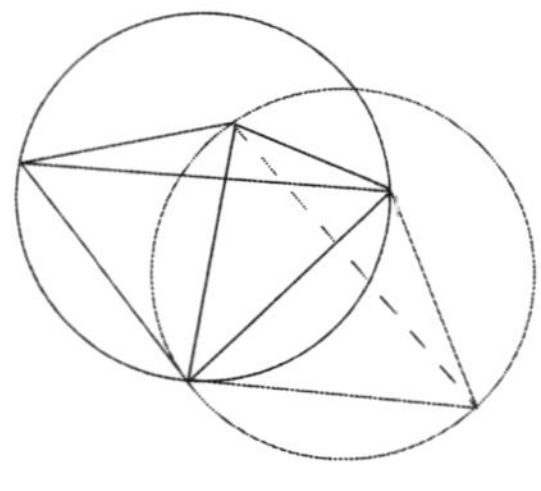

Figure 1: Delaunay tetrahedron.

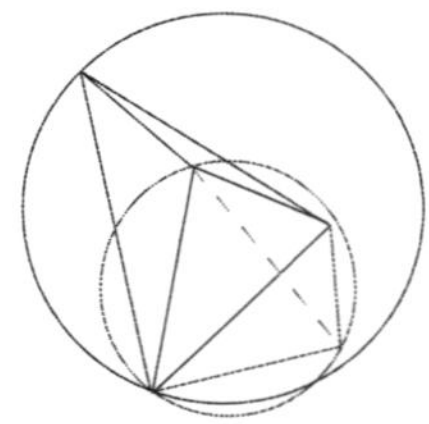

Figure 2: Not Delaunay tetrahedron.

elevation of river or bay, identification of geological boundary, etc. The finite element method has been successfully applied to solve a variety of engineering problems. On the other hand, it is dependent on the finite element mesh. The configuration of the objects to be analyzed are complicated, so that it is impossible to generate the finite element mesh by hand. Automatic mesh generation is necessary for the finite element method. The Delaunay method is an effective technique of mesh generation. They are introduced in following section.

2 Delaunay method

The Delaunay method is one of the technique to generate the finite element mesh. Tetrahedral elements are generated with the node group which is set arbitrarily in the computational domains. Circumscription sphere of a tetrahedron generated by the Delaunay method does not include nodes of other elements as shown in Figure 1. This is a feature of the Delaunay method. Figure 2 shows an example that is not the Delaunay tetrahedron. Using this method, a nearly regular tetrahedral mesh can be obtained. This geometry is in agreement with the preferable shape for the finite element method.

2.1 Creation of the tetrahedron using Super tetrahedron

A tetrahedron is known as the Super tetrahedron, if it includes a computational domain as shown in Figure 3. Super tetrahedron is useful for the computational domain into the tetrahedra. Sometimes, Super tetrahedron needs to change the size to take in the computational domain. Therefore, computational domain are covered by the tetrahedron. If a node is inserted in the computational domain, it must be in any tetrahedron. The first step of the Delaunay method is to divide the Super tetrahedron into four tetrahedra by the first node as shown in Figure 4. Basically, it is repetition that one element is divided into four tetrahedra by the inserted node.

2.2 Division of polyhedron

A node which is called P is set into the computational domain. If an element includes the node P is divided into four tetrahedra, the Delaunay method is not

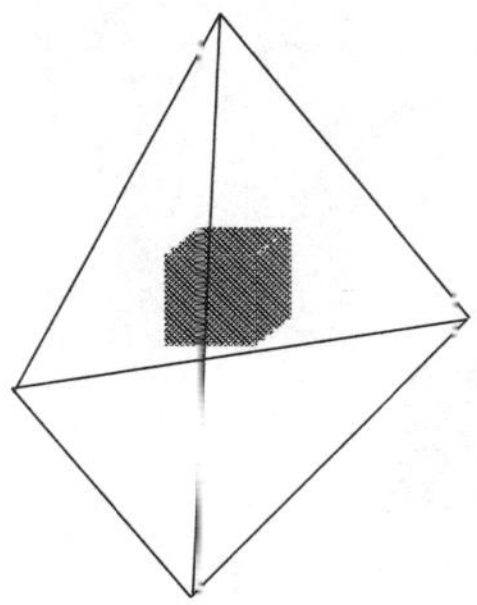

Figure 3: Super tetrahedron.

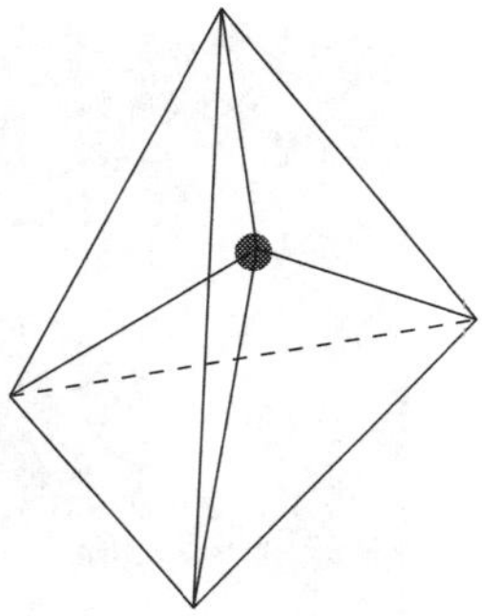

Figure 4: First step.

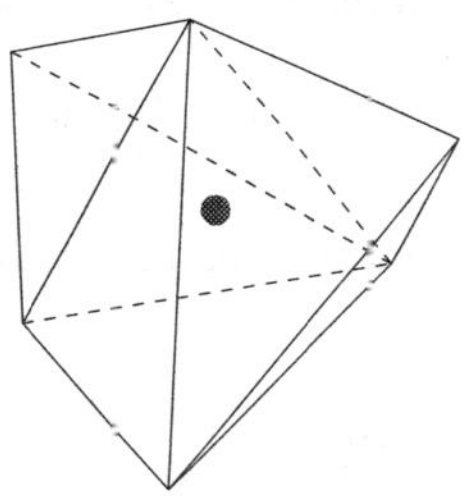

Figure 5: Polyhedron.

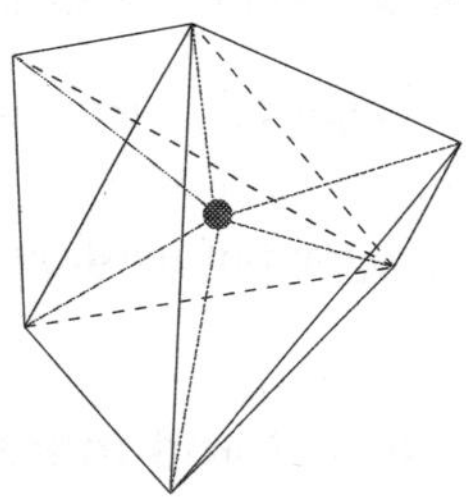

Figure 6: Generated mesh.

always satisfactory. If it is not satisfactory, these tetrahedra must be replaced according to the swapping algorithm. Elements which include the node P in the circumscription sphere of these elements are searched in the first step. Then, a polyhedron is generated with elements which are searched in the first step. If a polyhedron is generated, common sides of elements are removed as represented Figure 5. New elements are generated by connecting surfaces of the polyhedron and the node P as shown in Figure 6.

2.3 Algorithm of the Delaunay method

The algorithm of the Delaunay method is as follows.

1. Nodes data are input to the program.
2. Super tetrahedron is set to include a computational domain.
3. A node is inserted and search a tetrahedron which has a circumscription sphere including it.
4. Generate a polyhedron with tetrahedra which are found out.
5. Divide a polyhedron into tetrahedra.
6. If all points are inserted then Super tetrahedron is removed, else go to 3.

Figure 7: Suemune tunnel in Okayama, Japan.

A computer program of mesh generation is developed according to the above algorithm.

3 Suemune tunnel and mountain model

In this section, the Suemune tunnel model using the Virtual nodes is described. The Suemune tunnel is located in Okayama prefecture in Japan. To generate mountain and tunnel meshes, the Delaunay method is insufficient. The Delaunay method cannot dispose the complicated face. But, a lot of practical models have complicated faces. The practical model is not possible if the generation does not solve this problem. If generating practical model without taking measures, inadequate elements are generated at the tunnel mouth as shown in Figures 10 and 11. Many tricks are indispensable to generate various complicated shape. To clear up these problems, Virtual nodes are used.

3.1 Nodal distribution of surface and Suemune tunnel

From the provided data, nodal distribution of the tunnel is generated. The tunnels are often round in shape. To express roundness, the distance of each node becomes short. The surface of mountain and the tunnel mouth are close. In these areas, the distance between each node is incompatible. From the fine and coarse nodal distributions, the qualities of elements generated is low. Thus, it is better to subdivide the domain close to the mouth.

3.2 Virtual nodes

Figures 10 and 11 show inadequate elements at the tunnel mouth. These elements must be removed. Setting Virtual nodes on the tunnel mouth as shown in Figures 12 and 13, the distance is important. The Virtual nodes cannot exercise an

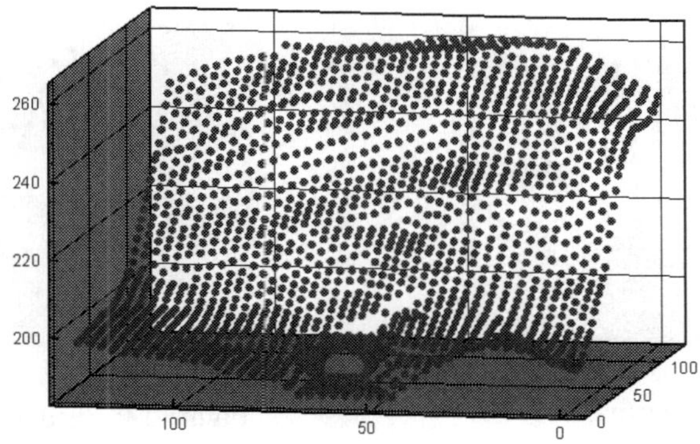

Figure 8: Nodes distributed on surface.

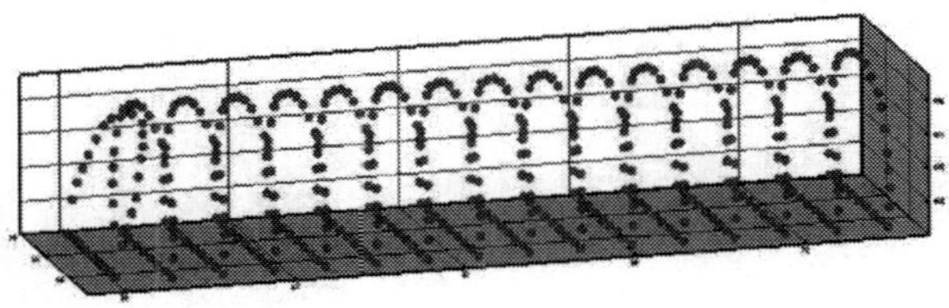

Figure 9: Nodes distributed for tunnel.

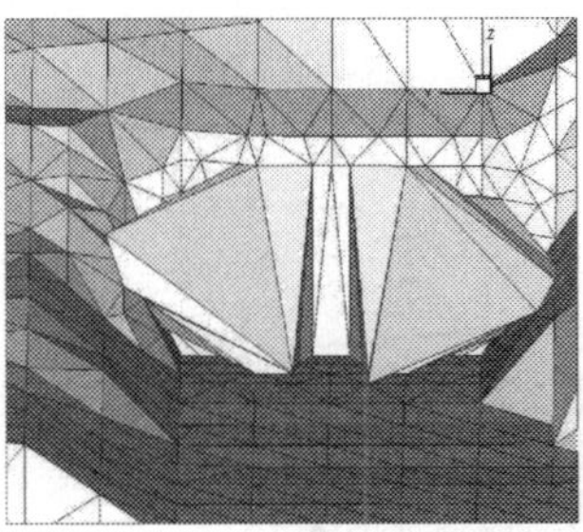

Figure 10: Inadequate elements 1.

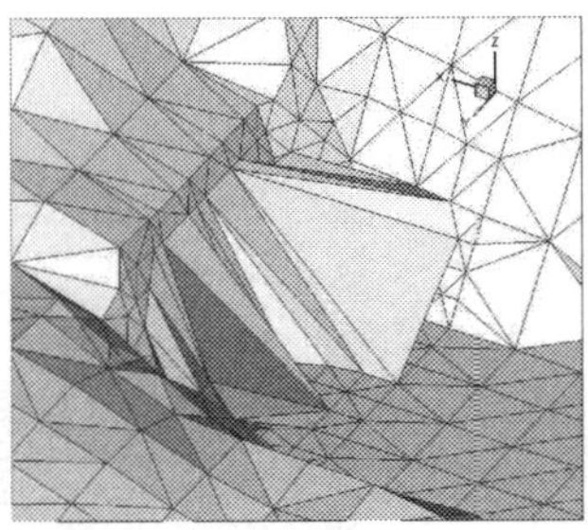

Figure 11: Inadequate elements 2.

effect if too far, or, too close. In this research, the distance between Virtual nodes is taken as an average of the interval of nodes. After the mesh generation, elements consisted of Virtual nodes are removed. Of course, Virtual nodes are removed as shown in Figures 14 and 15. The Virtual nodes are effective for not only surface but also interior boundaries. Virtual nodes and elements which consist of Virtual nodes are removed. Then, the finite element mesh which has interior boundaries is completed.

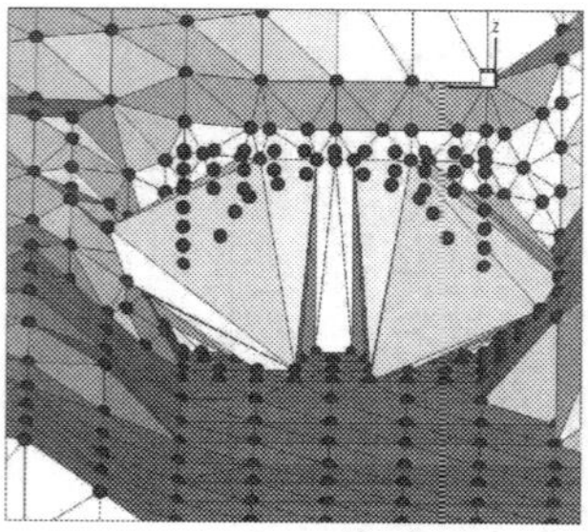

Figure 12: Virtual nodes 1.

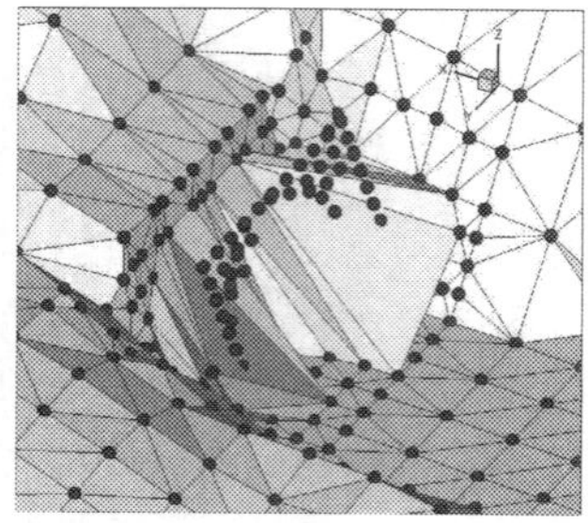

Figure 13: Virtual nodes 2.

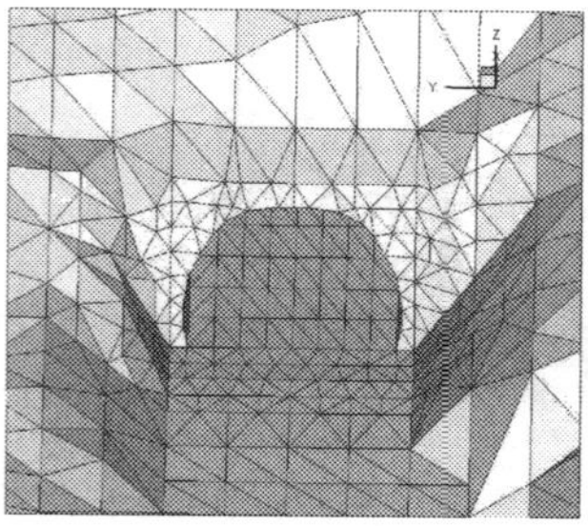

Figure 14: Tunnel mouth 1.

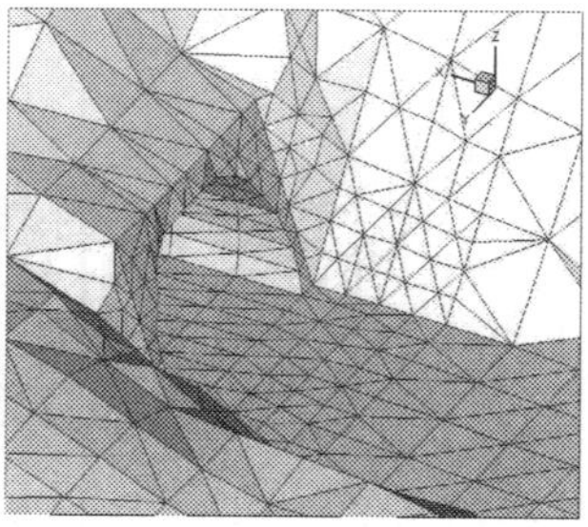

Figure 15: Tunnel mouth 2.

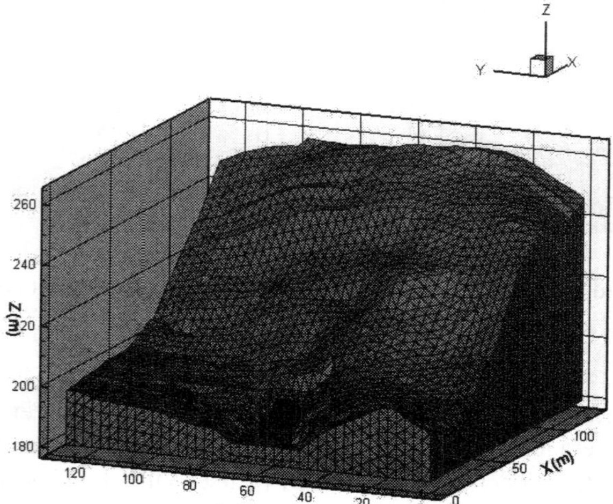

Figure 16: Completed mesh.

4 Conclusion

The 3-dimensional finite element mesh can be generated using the Delaunay method. But, the Delaunay method is not perfect in generating the numerical domain. There is a problem that the Delaunay method cannot recognize the concave face and interior boundaries, yet the problem was overcome using Virtual nodes. Then, the Delaunay method is possible to divide various complicated shape domain. The Virtual nodes are effective for complicated shape domain.

References

[1] George, P. L. and Hermeline, F. Delaunay's mesh of a convex polyhedron in dimension d. Application to arbitrary polyhedra. *International Journal for Numerical Methods in Engineering*, **33**, 975–995, 1992.

[2] Taniguchi, T. Automatic Mesh Generation for FEM, 1992.

[3] Weatherill, N. P. A method for generating irregular computational grids multiply connected planar domains. *International Journal for Numerical Methods in Engineering*, **8**, 181–197, 1988.

[4] Thompson, J. F. A general three-dimensional elliptic grid generation system on composite block structure. *Computer Methods in Applied Mechanics and Engineering* **64**, 377–411, 1987.

[5] Yerry, M. A. and Shephard, M. S. Automatic three-dimensional mesh generation by the modified-octree technique. *International Journal for Numerical Methods in Engineering*, **20**, 1965–1990, 1984.

[6] Watoson, D. F. Computing n-dimensional delaunay tessellation with application to voronoi polytopes. *The Computer Journal*, **24**(2), 167–172, 1981.

Emergency guidelines for two abandoned mines in Piani dei Resinelli area (Lecco)

M. Papini, L. Longoni & K. Dell'Orto
Politecnico di Milano, Italy

Abstract

The reconversion of abandoned mines for touristic aims tentatively began some years ago: now it has gained great acknowledgement from the public and specialists who are interested in the historic, scientific and cultural aspects of mines. Due to this increased presence of tourists it is necessary to ensure safety inside mines. In the last years the issue of safety in underground works aroused great attention in Italian, European and Extra-European state laws after the occurrence of serious lethal accidents inside road galleries, extractive mines and caves. However, these laws only apply to mines still in use, with have different problems than those occurring in abandoned mining sites. Old mines, now used for touristic purposes, are visited by people that don't know anything about problems that can occur in underground environments and don't perceive risk situations: they can display unpredictable behaviour and in critical conditions it is very difficult to keep the situation under control.

As a result, we developed these Emergency Guidelines (based on two mines) in order to create a model and originate general directives to be followed by other mines, as happens with other events in the Civil Protection field. First of all we defined risk scenarios and then traced the necessity of mobilization of operative structures. In addition, mines have been mapped with a GIS, including in the geo referenced map all the information about the mine that we thought to be useful to provide faster and more effective intervention in case of an emergency.

Keywords: abandoned mines, safety, GIS, risk scenarios.

1 Introduction

Safety inside mines is a point of fundamental importance, because of the greater probability of the occurrence of natural hazards in underground fields with respect to surface ones. In recent years the issue of safety in underground works has gained great attention in Italian, European and Extra-European state laws after the occurrence of serious mortal accidents inside road galleries, extractive mines and caves. But these laws apply only to operational mines, with different problems from those occurring in abandoned mining areas. In this way, safety in mines used for touristic aims has a priority to protect people who want adventurous, cultural or scientific experiences.

Until now the development of this new kind of tourism has continued without any law about safety and matters of Civil Protection, so in case of an accident there's no model to be followed to guarantee visitor safety. This emergency guideline, based on two mines, proposes to be a model for other similar visitor attractions, both in Italy and other countries. However every place has got specific features and some particulars may not been considered by this work in order to present it as a general guideline. In this way, the risk scenarios and intervention models presented below are made considering a particular mine but they can be used in a lot of other cases similar to the one presented.

2 The case study

Mining area containing Anna and Cavallo mines is located in Piani dei Resinelli, a locality of Abbadia Lariana (Lecco, Italy). In the past this area has been of great importance for the extraction of argentiferous galena, so a lot of mines have been opened to extract this mineral. It is possible to access to the Anna mine from rifugio SEL (Società Escursionistica Lecchese), at 1270m a.m.s.l., and following a short track along Val Calolden, reaching the principal gateway at 1220m a.m.s.l. (see the map in fig. 1). Access to Cavallo mine is possible starting from the 12th hairpin turn (road sign) long the road from Ballabio to Piani dei Resinelli. The area is reachable from Lecco city following the direction to Valsassina and then following signs to Piani dei Resinelli (from Lecco about 30–40 minutes by car).

Aid stations in the area are:

• **Lecco Hospital ("A. Manzoni")**	• **Fire Stations:**	• **Croce Rossa Italiana:**
	- Lecco headquarter station - Bellano station - Valmadrera station - Merate station	- Local Lecco committee - Valmadrera operative office - Ballabio office

To see the position of rescue stations in Lecco, see the map below (fig. 2):

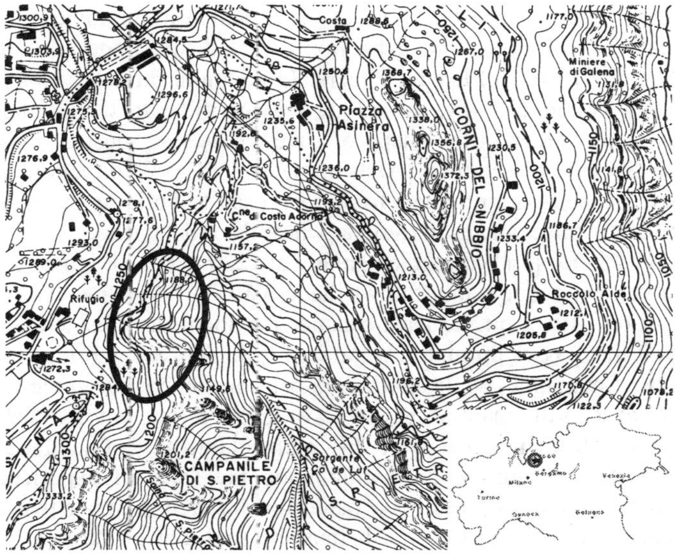

Figure 1: Map with position of Piani dei Resinelli Mine.

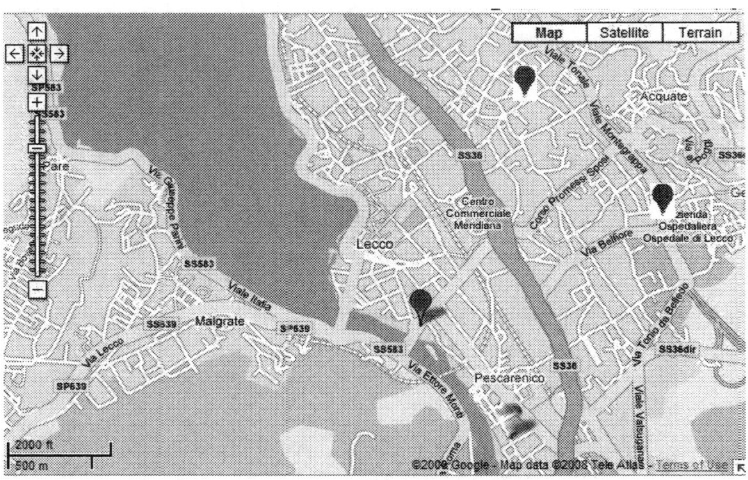

Figure 2: Position of rescue stations in Lecco: Fire Station (F), Hospital (H) and Croce Rossa (C) (map taken by Google Map, ©2008 Google).

2.1 Map of mines

The Anna mine map (the most visited mine in the area) has been made with the use of a terrestrial laser scanner, which generated a high-precision map. The Cavallo mine map (with a very low presence of tourists) instead has been made using analogical instruments (total station).

Over the centuries, different methods of extraction in the mines have created an irregular and complex plan; extraction by hand, dynamite and air hammer (Anna mine only) have all left their mark. In addition, mineral veins have been followed in a systematic way, sometimes leaving pillar to sustain the vault, so there's not a real order in the shape of caves; however, most of these tunnels follow the SW-NE direction in Anna mine and SE-NW direction in Cavallo mine. Some of the waste products have been used to build up stonewalls along most parts of the trace, and used to fill a part of the old cavities.

Ribasso Umberto is a "carreggio" gallery at 1220m a.m.s.l. linked to the Anna mine by a vertical gallery, and follows the WSW-ENE direction. The visit starts from the ticket office, on the square under the Piani dei Resinelli skyscraper. Then, a trail that passes through a wood into Calolden valley brings to the main access of the Anna mine, while another track brings the visitor to the 12th hairpin turn and from it to the main access of Cavallo mine. The whole track followed by tourists has been mapped with different colours to define three different routes suggested by guides, depending upon the level of experience of the visitors: route number one has been made for children, number two for adults and number three only for experts. It must be said that in the case of Cavallo mine only the expert route exists (the mine is characterized by extreme and wild conditions).

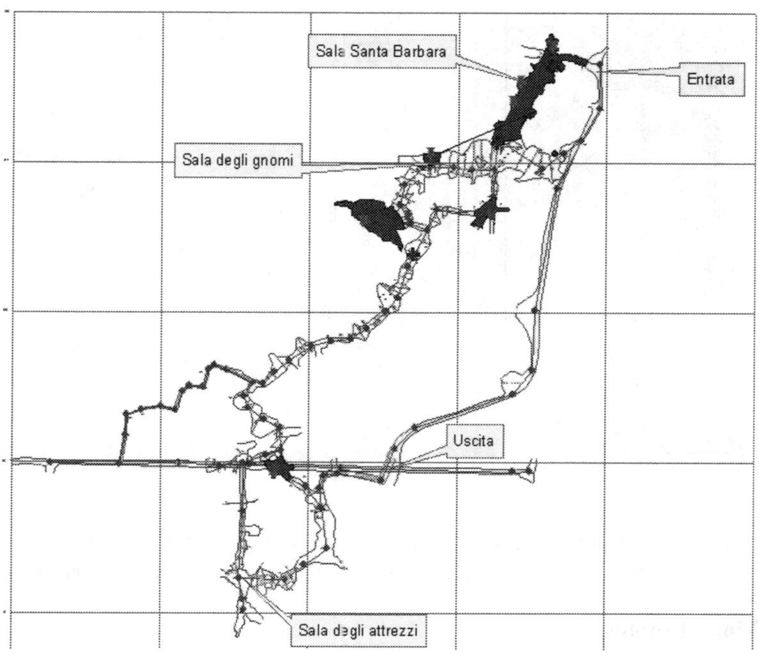

Figure 3: Map of Anna mine, with position of monitored areas (red) and crackmeters (blue).

During the visit, tourists pass through a route that, in general, lasts for 50 minutes, on a track that sometimes is slippery, often with a low vault (it is very easy to hit the head on the vault). For expert visitors it is also possible to visit the lower part of the Anna mine, Ribasso Umberto, which is accessible via a very steep stair. All tunnels are illuminated by a series of lights and emergency lights, although helmets with headlamps are also worn by tourists in order to have complete visibility; especially in Cavallo mine where the lights and emergency lights are mostly out of duty. From the ticket office a phone line runs to the Anna mine. There are two phones, one positioned at the beginning and one at the end of the mine track. All the tracks either in Anna and Cavallo mine are provided with two fire extinguishers and two first-aid kits.

In order to make the mine safer some tracks have been shored up with tree trunks. Principal cracks and faults are monitored with a system of crackmeters – either one dimensional or three dimensional. The position of the emergency exits, gathering areas and exit routes, position of phones, fire extinguishers, first aid kits and position of monitored areas, crackmeters and unstable blocks are described in the GIS maps (see an example in figs 3 and 4).

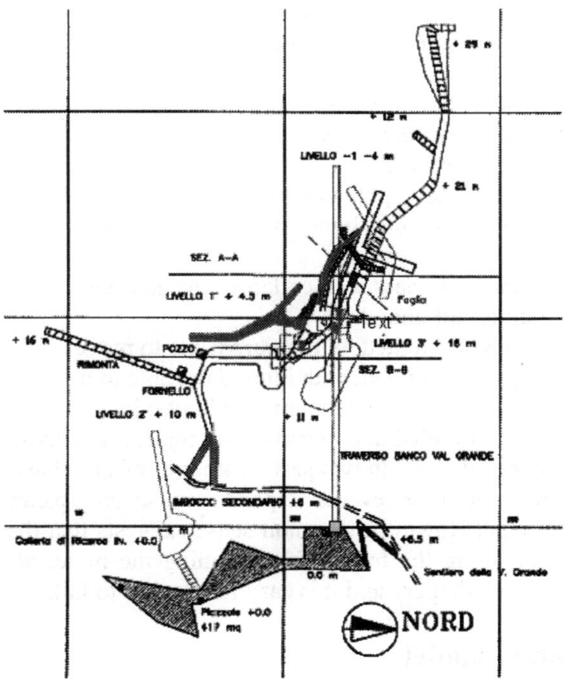

Figure 4: Map of Cavallo mine, with position of monitored areas (red) and crackmeters (blue).

3 Characterization of risks and risk scenarios

Risk events have been divided in two categories: external risks and internal risks. External risks concern injuries occurred during the walk from the ticket office to the mine (for example caused by the presence of mud on the walk).

Internal risks concern:

- Injuries to tourists due to sliding, hit of the head on the vault, etc.
- Damage to electric wiring.
- Collapse of rock blocks of every size and in every point inside the mine (however, it is important to notice that little collapses are more difficult to forecast than big ones).
- Collapse of rock walls inside the mine.

Risk scenarios are intended to be events that can occur inside and outside the mine, due to human and/or natural reasons. In the examined mine we identified several scenarios, with different probabilities of occurring:

1. Critical events outside the mine:

Injuries:

- The tourist is able move.
- The tourist is not able to move.

2. Critical events inside the mine:

Injuries:

- The tourist is able to move.
- The tourist is not able to move.

Damage of electric wiring:

- Start of emergency lights system.
- Damage of emergency lights system (emergency lights system doesn't start).

Instability of the rock mass with collapses inside caves:

- The group is still united.
- The group is still united, with injured able to move.
- The group is still united, with injured not able to move.
- The group is divided in two parts, without injured.
- The group is divided in two parts, with injured able to move.
- The group is divided in two parts, with injured not able to move.

The occurrence of these events depends on several parameters, so it is difficult to determine. However, common sense suggests that the most probable event is an injury along the trail before reaching the mine: after rainfalls the ground becomes very slippery and it is easy for tourists to fall.

4 Intervention model

The intervention model is an important point (perhaps the most important of the emergency guideline), and it offers a model on how to face emergencies, rescue people and restore normal state of life. First of all a figure is needed to be defined: the director of rescue operations, with the duty to work against the effects of critical events. In general, the Mayor of the city interested by the critical event, as the local Authority of Civil Protection activates emergency measures: as a support to the Mayor there is another figure, the Town Operative Representative (TOR), which has to manage staff, volunteers and emergency areas. The Mayor also establishes a Local Crisis Unity (LCU), made-up of people experienced in the field of emergency management.

In case of accidents inside the mine the settlement of LCU is not always done, it depends on the size of the event: it is assembled only when an event of significant dimensions occurs, like a collapse of vaults or any kind of relevant instability of the rock mass. In this case LCU is composed of:

- Abbadia Lariana Mayor
- TOR
- Comunità Montana del Lario Orientale (mine owner) President
- Mines safety representative (to be nominated by the mine Administrator)
- Technicians (to help in decisional, executive, administrative and technical phases)
- Local Police Office Commander
- Area Civil Protection Group manager.

This structure will be joined by other components (chosen by the Mayor): a Police officer, Firemen Chief, mine guides (supposing they know mines).

LCU components (available 24 hrs) will meet in an office as soon as possible in order to support the Mayor until the end of the emergency situation. For minor entity emergency scenarios, mine guides will elect between themselves a chief guide, as the Emergency Representative, to manage the emergency situation. Guides will also have to fill in a form (fig. 5) where they will register data about tourists in the mine.

It is important to mention the Emergency Coordinator, who is an expert of the mine, who has got the task to stay in the ticket office and to activate procedures that are necessary to pass out emergency.

Society	
Date:	**Weather:** □ **Calm Weather** □ **Changeable Weather** □ **Rainy Weather**
Start visit time	**End visit time hour ....**
Guide 1	**Guide 2 ....**
Number tourists .... **Typology** □ **School** □ **Children** □ **Adults** □ **Adults + Children**	**Given equipment:** ▪ **Number lights ....** ▪ **Number helmets................................** ▪ **Other:**
Trail: □ **Standard** □ **Reduced** □ **Standard + Ribasso Umberto**	**Notes:**

Figure 5: Form to be filled in for every visit to mines.

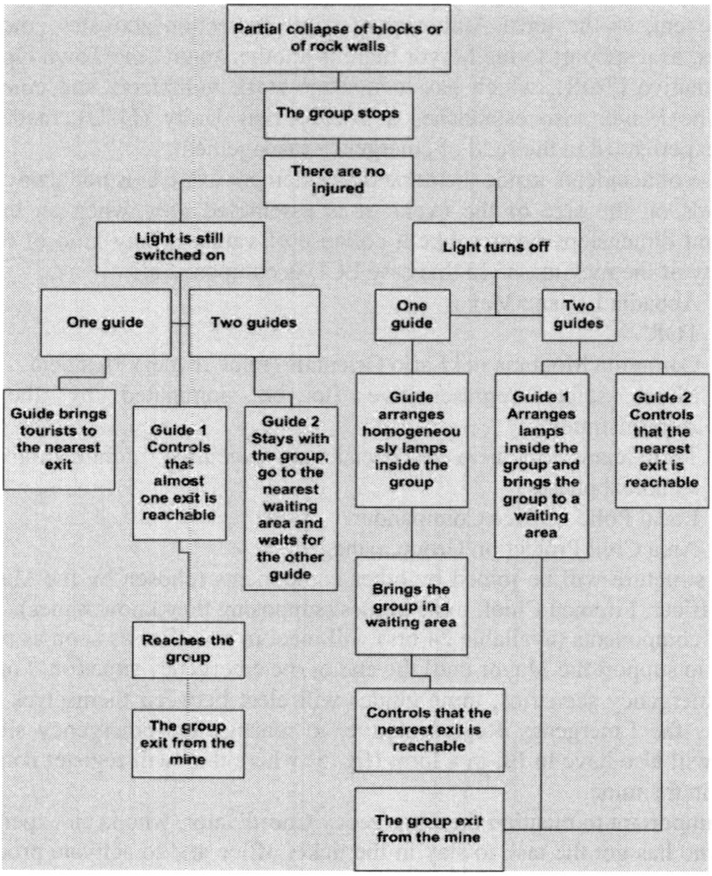

Figure 6: Scheme of an intervention model for a risk scenario (in this case, collapse of rock walls).

The Emergency Representative together with Emergency Coordinator and Safety Representative will activate emergency procedures and contact Civil Protection institutions (Firemen, Police, etc.).

Procedures to be followed are schemed with three steps:

1. Events tree
2. Actions tree
3. Intervention tree

However it is too difficult to singularly schematize these three trees, so we considered it to be better to make a scheme composed of a singular tree that includes all aspects together: events, actions and interventions (see example in fig. 6).

In order to describe the sequence of procedures to be followed for every considered scenario it is better to define “waiting areas” inside the mine (places where to bring tourist in case of an accident) and emergency exits.

Waiting areas have to be safe and wide, in order to contain at least 30 people
Intervention practices must be divided following a sequence of codes:

1. Pre-alarm code.
2. Alarm code.
3. Emergency code.

As events that could happen inside the mine are difficult to forecast, it would be better to be always ready for the emergency code. However, if during the monitoring of crackmeters an operator detects a movement, safety representative activates the alarm code and the mine is closed until the cause of the movement has been detected and removed. The succession of the events is schematized in tree diagrams in order to give easy access to intervention modalities and a fast check of the guideline.

5 Application of a Geographic Information System (GIS)

A GIS (Geographic Information System) is a software instrument useful to acquire, store, extract, transform and visualize spatial data from the real world. This system allows for the production, management and analysis of spatial data linking every geographic element with one or more alpha numeric descriptions.

Using it for mapping mines is the best way to obtain a clear representation of the mine and of all points of the map considered to be interesting for the management of an emergency guideline: location of lights, emergency lights, emergency exits, emergency areas, phones, first aid kits, etc. Every point has to be provided with a description of its characteristics and with a link to its photo.

Maps for this guideline have been intended as something interactive, in order to allow everyone to quickly acquire information on the mine: fast data retrieval is fundamental in case of an emergency situation, where loss of time has never to be considered.

6 Conclusions

The aim of this guideline is to safeguard people visiting mines, analyzing all possible risk scenarios (and the probability of each one happening) in order to be able to manage every kind of accident that can occur during a visit. The only case that has not been considered is a fire inside the mine, as the probability of that happening is very low (the humidity inside mines is very high and wood used to sustain vaults is often wet).

In order to validate an emergency guideline system it is fundamental to execute a text exercise, to find possible bugs and to allow it to be updated with eventual new parts. So at least one exercise per year has to be done, simulating a case of accident to test how the system runs.

It is important to say that from the analysis of possible scenarios it resulted that for every visit the presence of at least two guides in the mine is necessary: in case of an emergency the first one would stay with tourists and the other would go out and look for help. In particular, the presence of two guides is highly recommended for visits made after a long rainy period: the ground inside mines gets very slick and the probability of low entity collapses is high.

References

[1] Giussani Alberto, Relazione sul rilievo topografico delle miniere "Anna" e del "Ribasso Umberto" ai Piani dei Resinelli, Politecnico di Milano, 2006.
[2] Rodeghiero F., Jadoul F., Vailati G. e Venerandi I. (1986), Dati preliminari sulle mineralizzazioni a Pb-Zn dell'area tra Mandello e Ballabio (Lombardia Centrale). Memorie della Società Geologica Italiana, 32: 133-150.
[3] Papini Monica et al., Caratterizzazione geomeccanica con metodi statistici e valutazione MRES del rischio geologico presso i cantieri della Miniera Anna e del Ribasso Umberto, Politecnico di Milano, 2006.
[4] Papini et al., Requisiti e linee generali per la realizzazione di un sistema di monitoraggio delle miniere dei Piani dei Resinelli, Politecnico di Milano, 2007.

Damage zones near excavations: plastic solution by means of stress trajectories

P. Haderka & A. N. Galybin
Wessex Institute of Technology, Southampton, UK

Abstract

This study presents an alternative approach for identification of damaged zones near excavations. The approach is based on ideal plastic solutions but in contrast to the classical case it deals with the Cauchy's problem only by alternating classical solutions for slip zones with solutions for stress trajectories followed by conversion of the latter into slip grids. Comparisons with classical solutions are discussed.

Keywords: ideal plasticity, slip lines, stress trajectories.

1 Introduction

Damaged zones near excavations are frequently observed in underground mines. They present a significant issue for safety and effectiveness of mining operations and should be properly accounted for in design of particular excavations. This paper is aimed at the development of numerical methods for determination of damage zones in the case of long-wall excavations such as tunnels, well-bores, tabular stopes, etc. Another important motivation of this study is the application of a variant of the stress trajectory element method, STEM, which is currently under development in the Wessex Institute of Technology [1]. The method addresses the problem of stress identification in statically determined bodies by employing stress trajectories.

The concept of stress trajectories comes from photoelasticity, therefore one can adopt the following definition due to Frocht [2]: Stress trajectories are curves the tangents to which represent the directions of one of the principal stresses at the points of tangency. Stress is a second-rank tensor which components satisfy differential equations of equilibrium, DEE, and certain constitutive equations. The latter constitute a broad class and some examples are found in engineering:

- in elasticity, the laplacian applied to the first invariant of the stress tensor should vanish;
- in ideal plasticity, the deviator of the stress tensor is a constant;
- in granular medium, certain linear relationships between the mean stresses and the stress deviator should be fulfilled;
- in rock mechanics non-linear relationships are frequently used.

This study employs the stress trajectory concept for the case of ideal plasticity in order to build slip grids in plastic zones forming near an excavation.

2 Mathematical model of damaged zones near long-wall excavations

2.1 Assumptions in modelling

It is accepted that stress field near an excavation can be described by the plane stress condition. This is typical for tunnels, well-bores and other long-wall excavations provided that their axes are oriented along one of the principal stress. It is assumed that the excavation has been made in weak rocks that can be deformed beyond the yielding limit and therefore surrounding rocks can be damaged. It is also assumed that the rockmass in damage zones is in limiting equilibrium. For simplicity, the damage zones are considered to be ideal plastic. Such an assumption is often accepted for coals, e.g. [3], however this simplification is not crucial and can be easily extended for some other rocks staying in limiting equilibrium to take friction and cohesion into account, see [4].The plastic zone is assumed to be bounded and embedded into infinite elastic zone. The boundary between the zones is not known. The present paper does not deal with the identification of the unknown boundary; if necessary this can be performed by the approach suggested in [5].

2.2 Lamé – Maxwell equations of equilibrium

For plane problems there are two independent DEE valid at each point of the domain including the boundary between plastic and elastic zones.

$$\frac{\partial\sigma_{11}}{\partial x_1}+\frac{\partial\sigma_{12}}{\partial x_2}=0,\quad \frac{\partial\sigma_{12}}{\partial x_1}+\frac{\partial\sigma_{22}}{\partial x_2}=0 \tag{1}$$

where σ_{kj} are stress components in a Cartesian coordinate system Ox_1x_2. The gravitational term has been omitted in (1) for convenience, because it can be accounted for by incorporation into boundary conditions.

DEE (1) can be rewritten in terms of principal stresses and principal directions in the Lame-Maxwell form, [3]:

$$\frac{\partial\sigma_1}{\partial s_1}+\frac{\sigma_1-\sigma_2}{\rho_2}=0,\quad \frac{\partial\sigma_2}{\partial s_2}+\frac{\sigma_1-\sigma_2}{\rho_1}=0 \tag{2}$$

Here σ_k is a principal stress along the k-th stress trajectory which radius of curvature is defined as $\rho_k^{-1}=\partial\theta/\partial s_k$; θ is the principal direction (the angle between σ_1 and the x_1-axis); s_k is the arc length along the k-th trajectory. In order to distinguish different families of the stress trajectories it is assumed that $\sigma_2 \leq \sigma_1$, these families are further referred to as s_1 and s_2 – families.

In plastic zone the following governing equation is valid

$$\frac{\sigma_1 - \sigma_2}{2} = k \tag{3}$$

where k denotes the yielding limit of the rocks.

The systems (1), (3) or (2), (3) form closed systems of equations and can be solved independently of kinematics equations. The latter is not included in the analysis because it is not required for the statically determined cases as the one considered here.

2.3 Classical approach

Classical texts on ideal plasticity, e.g. [6, 7], suggest to rewrite (1) by substituting (3) and using the following relationships

$$\sigma_{11} = \sigma - k\sin(2\vartheta), \quad \sigma_{22} = \sigma + k\sin(2\vartheta), \quad \sigma_{12} = k\cos(2\vartheta) \tag{4}$$

where the angle

$$\vartheta = \theta - \pi/4 \tag{5}$$

represents the inclination of the s_α-family of slip lines; another orthogonal family, the s_β-family, has orientations $\theta + \pi/4$. Notations are shown in Fig. 1. This leads to the system of 2 PDE with respect to 2 unknowns σ and ϑ:

$$\begin{aligned} &\frac{\partial\sigma}{\partial x_1} - 2k\left[\cos(2\vartheta)\frac{\partial\vartheta}{\partial x_1} + \sin(2\vartheta)\frac{\partial\vartheta}{\partial x_2}\right] = 0 \\ &\frac{\partial\sigma}{\partial x_2} - 2k\left[\sin(2\vartheta)\frac{\partial\vartheta}{\partial x_1} - \cos(2\vartheta)\frac{\partial\vartheta}{\partial x_2}\right] = 0 \end{aligned} \tag{6}$$

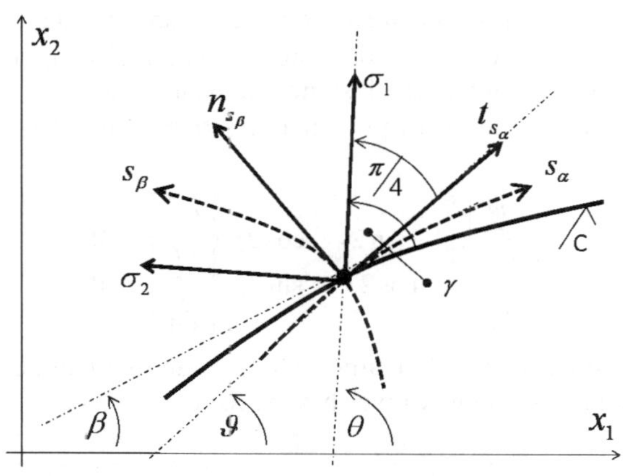

Figure 1: Notations: β is the angle of the tangent to the contour C with x_1 axis, γ is the angle between the tangent to the contour and the orientation of the first principal stress σ_1. $t_{s\alpha}$ and $n_{s\alpha}$ show the tangent and normal to the slip lines s_α and s_β.

After simple transformation of (6) one finds the Henckey integrals and differential equations for determination of the slip grid [5]

$$\frac{dx_2}{dx_1} = \tan\vartheta, \quad \frac{\sigma}{2k} - \vartheta = \text{const}, \quad \text{for} \quad \alpha-\text{family};$$
$$\frac{dx_2}{dx_1} = -\cot\vartheta, \quad \frac{\sigma}{2k} + \vartheta = \text{const}, \quad \text{for} \quad \beta-\text{family}; \tag{7}$$

After that (7) is solved by using the finite difference method.

The classical Cauchy's problem is solved within a characteristic triangle which boundaries are formed from the contour as the hypotenuse and characteristic lines as its legs. The legs of the triangle can be taken as boundaries for solving the Riemann – Goursat problem providing the solution when boundaries are characteristic lines of different families. By solving this problem the integral surface can be considerably extended.

2.4 STEM approach

STEM solves the Lame-Maxwell DEE's in the following form

$$\frac{\partial\psi}{\partial s_1} + \frac{\partial\theta}{\partial s_2} = 0 \quad ; \quad \frac{\partial\psi}{\partial s_2} + \frac{\partial\theta}{\partial s_1} = 0 \tag{8}$$

where $\psi = \sigma/2k$.

Its evident that the 2nd order PDE has the form

$$\frac{\partial^2\theta}{\partial s_1^2} - \frac{\partial^2\theta}{\partial s_2^2} = 0 \tag{9}$$

It should be noted that mean stresses (σ and ψ) and the angle ϑ also satisfy this equation. This forms the basis for using alternating Cauchy problems for the slip grids and stress trajectories as illustrated in the next section.

To formulate BVP in terms of principal direction, θ and derivatives of θ are needed

$$\begin{bmatrix} \dfrac{\partial\psi}{\partial t} \\ \dfrac{\partial\psi}{\partial n} \end{bmatrix} + \begin{bmatrix} -\sin 2\gamma & \cos 2\gamma \\ \cos 2\gamma & \sin 2\gamma \end{bmatrix} \begin{bmatrix} \dfrac{\partial\theta}{\partial t} \\ \dfrac{\partial\theta}{\partial n} \end{bmatrix} = \begin{bmatrix} 0 \\ 0 \end{bmatrix} \tag{10}$$

The decomposition of the derivatives along s_1 and s_2–families from (8) into their normal and tangential derivatives has the form

$$\frac{\partial}{\partial s_1} = \cos\gamma\frac{\partial}{\partial t} + \sin\gamma\frac{\partial}{\partial n} \quad , \quad \frac{\partial}{\partial s_2} = -\sin\gamma\frac{\partial}{\partial t} + \cos\gamma\frac{\partial}{\partial n} \tag{11}$$

Using equation (10) the unknown normal derivatives of stresses and orientations can be calculated.

Nodes of the next layers are found using the properties of characteristics, one of which states that characteristics form an orthogonal grid; characteristics of different families form an orthogonal system of lines. Intersections of these different families create a set of nodes for the next layer of the grid.

$$z_j^{k+1} = z_j^k + \delta_j^{s_1} e^{i\theta_j^k} \quad , \quad z_j^{k+1} = z_{j+1}^k + \delta_{j+1}^{s_2} e^{i\left(\theta_{j+1}^k + \frac{\pi}{2}\right)} \tag{12}$$

Notations are shown in Fig. 2. In formulas (12) the subscript refers to the node while the superscript specifies the layer number (layer k is known, layer $k+1$ is unknown); δ_j represents the distance between nodes of the layer k and $k+1$ at position j or $j+1$ along the s_1 or s_2 characteristic.

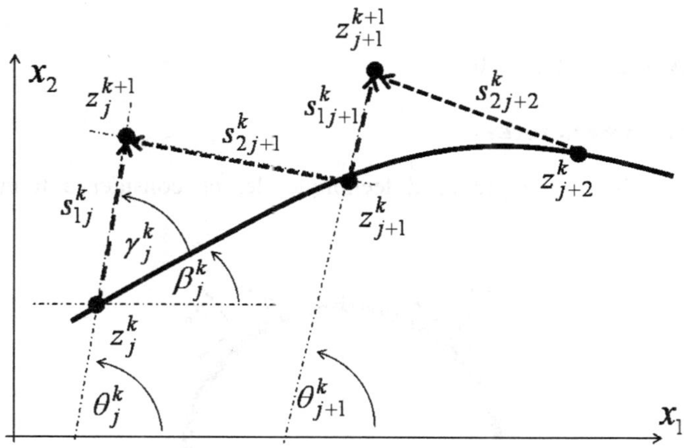

Figure 2: Determination of coordinates and visualization of angles.

At this point the normal derivatives of orientations and stresses as well as the tangential ones are determined and the change of the orientations and stresses along characteristics can be computed using (11). These terms are essential for computing the unknown values in the new layers, for which the Taylor s expansion was used

$$\theta_j^{k+1} = \theta_j^k + \left|z_j^{k+1} - z_j^k\right| \cdot \left.\frac{\partial \theta}{\partial s_1}\right|_j^k \quad , \quad \psi_j^{k+1} = \psi_j^k + \left|z_j^{k+1} - z_j^k\right| \cdot \left.\frac{\partial \psi}{\partial s_1}\right|_j^k \tag{13}$$

2.5 Combining STEM and Kachanov

Classical approach developed for example by Kachanov [7] gives the solution only inside the characteristic triangle stating the solution outside cannot be obtained.

Boundary conditions from the classical approach can be used in STEM to demonstrate the equivalence between these two approaches. Slip Grid (SG) is obtained by using classical solution of Cauchy's problem for the given non-characteristic boundary while Stress Trajectories (ST) for the corresponding boundary are found by STEM. By rotating the stress orientations in the ST Grid by $-\pi/4$, orientations of shear stresses are found at every node.

Solution of the Cauchy problem using STEM gives a characteristic triangle again, which is different from that obtained in the classical approach. The boundaries of this triangle can be used in the next step as new initial boundaries. Bearing in mind that ST Grid was obtained using STEM the following needs to be brought in attention. Boundaries of the triangle are characteristic lines which is why they would form a characteristic boundary for ST. Rotating the orientations on this boundary by $-\pi/4$ forms a non-characteristic boundary for the slip lines. This new boundary can be then easily used for classical solution.

These two approaches can be alternated to move away from the initial boundary.

3 Numerical examples

3.1 Tunnel in weak rocks

As an example of the proposed technique, let us consider a tunnel of the geometry shown in Fig. 3.

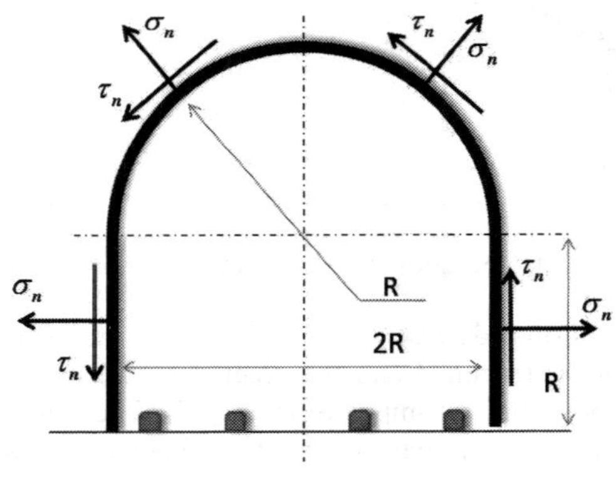

Figure 3: Tunnel in weak rocks (σ_n and τ_n are present to model the effects of the supports).

The BC are chosen as follows

$$\sigma_n = c \quad , \quad \tau_n(s) = as + b \tag{11}$$

where *a b* and *c* are certain piecewise constant coefficients different from linear and circular parts of the boundary.

3.2 Results

The following results have been obtained for the classical solution of Cauchy's problem (using Kachanov's approach); solution by STEM; comparison of STEM approach and classical approach. Due to symmetry the problem has been solved for the left half of tunnel's boundary as shown in Fig. 3.

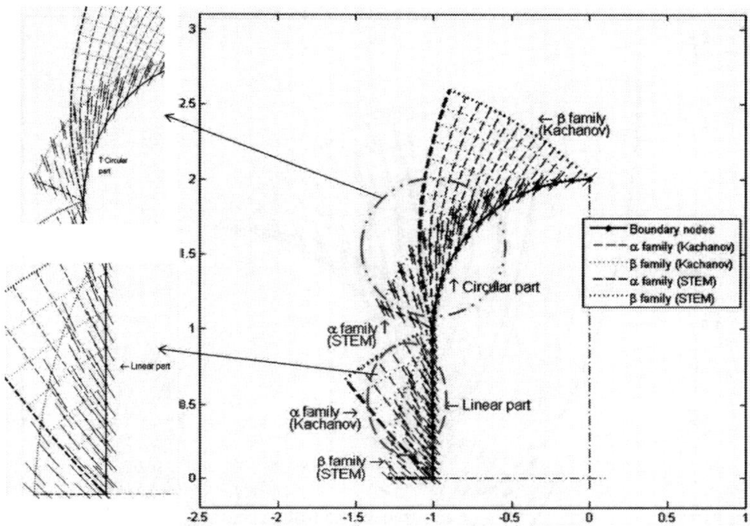

Figure 4: Classical solution (Slip Grid) vs. STEM solution (ST Grid).

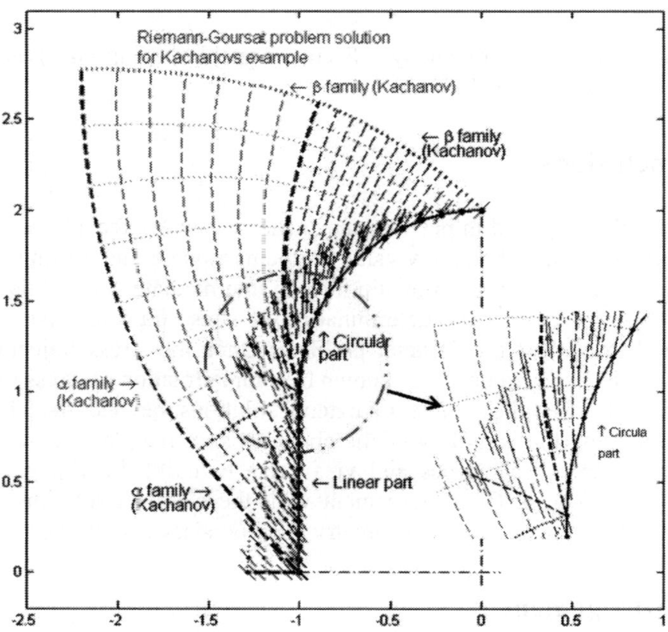

Figure 5: Extended solution using Riemann – Goursat problem (Slip Grid) versus STEM solution (ST Grid).

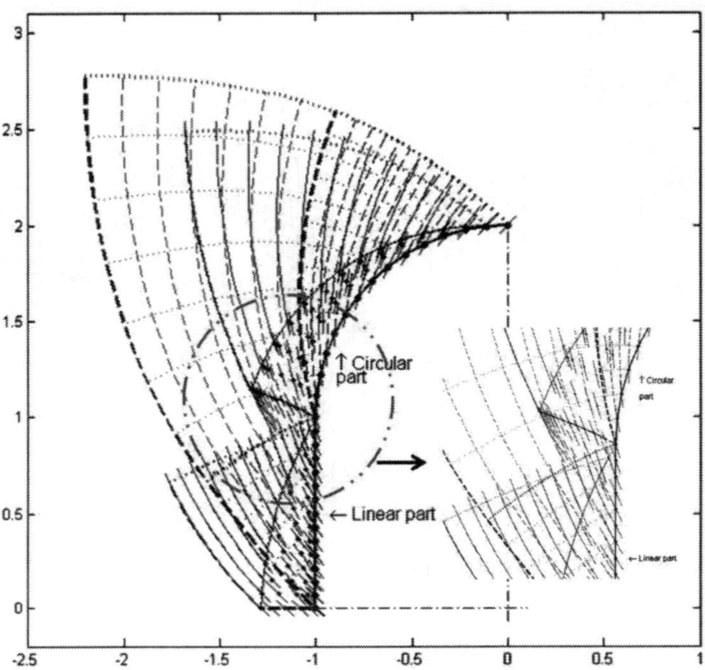

Figure 6: Solution of Cauchy – Riemann problem (Slip Grid) and extended solution by STEM.

4 Conclusions

It is shown that slip grid in plastic zones near long-wall excavations can be built by solving the Cauchy boundary value problems alone without using the Goursat problem for the determination of slip lines in intermediate zones. It is shown that the Cauchy problem for the determination of stress trajectory patterns can be formulated instead of the Goursat problem. Once the stress trajectory field is determined the slip grid becomes known by simple rotation of stress orientations. The approach also allows one to extend solutions beyond the characteristic triangle by using the boundary of the characteristic triangle as a new boundary for building stress trajectories and visa versa to build the slip grid from the boundary of ST grid. Good agreement with the classical solutions have been observed for particular cases of geometry and boundary conditions.

Acknowledgement

The authors acknowledge the support of EPSRC through Research Grant EP/E032494/1.

References

[1] Galybin, A.N. 2007. Introduction of STEM for stress analysis in statically determined bodies. *WIT Transactions on Modelling and Simulation*, **44**, WIT Press, Southampton, UK, 79–88;

[2] Frocht, M.M. Photoelasticity, Vol. 1. Wiley, New York, 1941;

[3] Khistianovich S.A., Salganik, R.L., 1983; Several basic aspects of the forming of sudden outbursts of coal (rock) and gas; 5th Congress International Congress on Rock Mechanics, Melbourne; pp. E41–E50

[4] Goodman, R.E. Introduction to Rock Mechanics. J Wiley, New York, 1980.

[5] Cherepanov, G. P. 1963. On a method of solving the elasto-plastic problem Journal of Applied Mathematics and Mechanics, 27 (3), 644–655

[6] Hill, R.; The Mathematical Theory of Plasticity; Oxford, Clarendon Press; 1950;

[7] Kachanov, L.M.; Fundamentals of Theory of Plasticity; Published: Foundations of the theory of plasticity, Amsterdam, 1971;

CFD simulation of aerodynamic resistance in underground spaces ventilation

I. Diego, S. Torno & J. Toraño
GIMOC, Mining Engineering and Civil Works Research Group, Oviedo School of Mines, University of Oviedo, Independencia 13, 33004 Oviedo, Spain

Abstract

The main problems once someone gets deep inside an underground space are the lack of light and air. If the underground space is long enough or there are pollutants or heat sources inside the space the need for a ventilation system becomes essential. In the particular case of road tunnels, the need to dissipate engines fumes also raises the need to account for the high risk present of fire inside the tunnels. It seems clear that whatever the underground space is designed for, a ventilation system must be present during the construction and subsequent operative phase of the installation. A state of the art tool to calculate the ventilation system is computational fluid dynamics (CFD). The air in the underground space is discretized in finite volumes and mathematical methods are used to obtain the pressure and velocity fields all over the domain, capturing and analyzing all possible flow details, no matter the geometry if the mesh is fine enough. One of the main goals in classical ventilation calculations is to obtain the overall pressure drop of the air as it passes through the cavity, as this will guide the sizing and selection of the installation fan(s). This paper uses the commercial CFD code Ansys CFX 10.0 to calculate the pressure losses in a tunnel installation, comparing the results with the classical frictional losses calculations procedure. These simulations will guide a methodology of using CFD to calculate sections of the underground ventilation or even, if mesh sizes and computer means are big enough, to fully calculate ventilation over all domains. These studies have been carried out in the framework of the Research Project CTM2005-00187/TECNO, "Prediction models and prevention systems in the particle atmospheric contamination in an industrial environment" of the Spanish National R+D Plan of the Ministry of Education and Science, 2004–2007 period.

Keywords: CFD, turbulence, tunnel resistance, pressure drop.

WIT Transactions on the Built Environment, Vol 102, © 2008 WIT Press
www.witpress.com, ISSN 1743-3509 (on-line)
doi:10.2495/US080131

1 Introduction

The research group of the Mining Engineering and Civil Works of the University of Oviedo, based in the School of Mines of Oviedo (Spain) is developing a research project granted by the Spanish Ministry of Education and Science. One of the project goals is to develop and study dust emission factors in several industrial situations, from quarry blasts to ship unloading or loading facilities. The researches and simulations have been very successful in several fields obtaining interesting results in the case of dust coming from quarry blasts, both using computational fluid dynamics (CFD) [1] and classical dispersion methods, [2]. The research group has also experienced simulating air flows in underground spaces ventilated through auxiliary systems [3].

CFD seems to be the perfect tool to calculate air flows in underground spaces, both the velocity fields [4, 5] and the pressure fields [6, 7], with immediate applications in HVAC design [8] and security regarding fires and smoke [9].

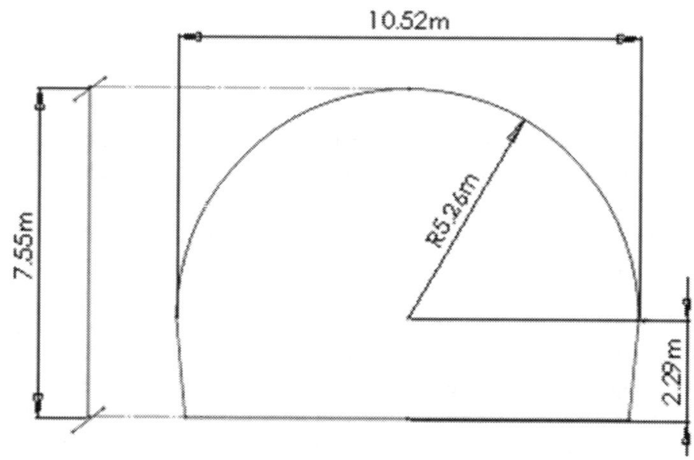

Figure 1: Tunnel section.

The next step to be taken is to translate the dust dispersion calculations to underground environments. In order to check the accuracy of the CFD simulations the pressure drop calculated by the CFD methods will be compared to the results of the classical frictional methods based in Darcy-Weisbach formulations. An assumption is made: if the pressure drop calculation is correctly developed by the CFD model then its accuracy would be proved and then the calculated velocity/pressure fields will be used as a base for multiphase simulations involving species other than air.

In this paper a section of 500 meters of a tunnel, Figure 1, will be simulated and the flow through it simulated. Its cross section is 67.07 square meters and its perimeter is 31.22 meters.

2 Theoretical head losses

To calculate the head losses created by turbulent flows in non circular ducts the modified Darcy-Weisbach equation [10] is used:

$$\Delta_{p,f} = \frac{1}{2} \cdot \frac{f \cdot L \cdot V^2 \cdot \rho}{D_H} \tag{1}$$

where
Δp,f: fictional head loss (Pa)
V: velocity of the air flow (m/s)
f: friction factor
ρ: air density (kg/m^3)
D: hydraulic diameter of the duct (m)
L: length of the duct (m)

In order to calculate f there are several empirical equations that relates f to Reynolds number and to the roughness of the wall. There will be used the Colebrook-White one [10]:

$$\frac{1}{\sqrt{f}} = -2\log\left(\frac{\varepsilon_r}{3.7} + \frac{2.51}{Re\sqrt{f}}\right) \tag{2}$$

where
f: friction factor
ε_r: relative roughness = ε/D.
Re: Reynolds number

$$Re = \frac{v \cdot D_H}{\upsilon} \tag{3}$$

where
v: velocity (m/s)
D_H: hydraulic diameter (m)
n: cinematic viscosity (m^2/s)

The values of the tunnel roughness are taken from [11] as reflected in table 1.

3 CFD simulation

The equations that govern the fluid flow are the Navier Stokes, eqn (4), which relates the velocity and pressure fields as well as density; the continuity equation, eqn (5), that express the mass conservation; and finally the energy equation, eqn (6), which relates also the temperature fields. These expressions create a system of differential equations that can only be solved, in the vast majority of cases, by numerical methods.

$$\rho \frac{D\overline{V}}{Dt} = -\overline{\nabla} p + \rho \overline{g} + \mu \nabla^2 \overline{V} \tag{4}$$

$$\frac{D\rho}{Dt} + \rho\overline{\nabla} \bullet \overline{V} = 0 \quad (5)$$

$$\rho\frac{D\widetilde{u}}{Dt} = K\nabla^2 T - p\overline{\nabla} \bullet \overline{V} \quad (6)$$

where

ρ: Density	∇: Gradient
$\nabla \bullet$: divergence operator	**t**: Time
U: velocity vector	**V**: velocity
p: pressure	**g**: gravity
μ: viscosity	$\widetilde{u}$: specific heat
T: Temperature	**K**: conductivity

Table 1: Tunnel roughness.

Tunnel support system	**Roughness (mm)**
Tunnel lining by shotcrete	
Very smooth surface	0.3
Medium conditions	2.5
Rough surface	9
Tunnel lining by concrete	2.5
Tunnel excavated in rock and no lining	
roughness surface	100
high roughness surface	200
irregular surface	300

In turbulent flow there have also to be solved additional equations that allow the calculation of the velocity and pressure fields in all the domains, the so called "turbulence models". Taking into account just the RANS (Reynolds Averaged Navier Stokes) they vary from the more or less simple where just one equation is added to the calculations, as the Spalart-Allmaras model, to the two equations models, k-epsilon or even seven equations, Shear Stress Transport (SST) models. This paper will show calculations that use k-epsilon models and scalable wall transport functions, which simplify the meshes used in the vicinity of the walls. All the calculations will be made using commercial code Ansys CFX 10.0. The domain to be calculated is divided in finite volumes where the former equations will be solved by linear methods. A base in this methodology is to demonstrate the independency of the results to the type and size of the mesh.

The mesh is done in three dimensions in a non structured way, using tetrahedral, prismatic or pyramidal elements developed using software ICEM CFD 10.0 starting from parametric geometrical models developed in

SolidWorks. Tetrahedrons and pyramids cover the most part of the domain, but the prisms are used to gather calculation nodes in the vicinity of the wall, where the velocity gradient is high. Figure 2 shows a cross section and a longitudinal section of the mesh, whereas figure 3 shows details of prism layers close to wall and ground.

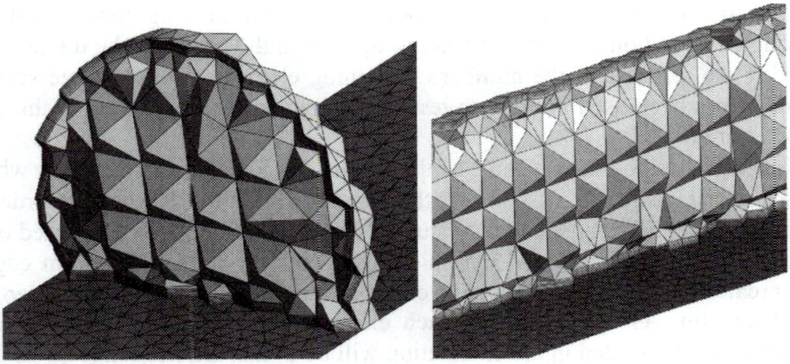

Figure 2: Cross section (left) and longitudinal section (right) of mesh.

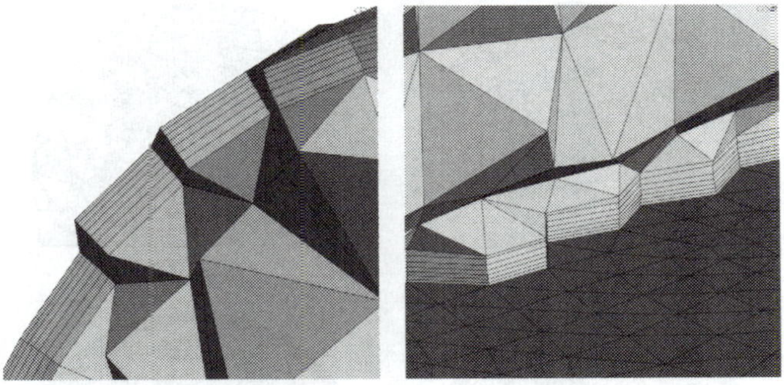

Figure 3: Details of mesh in the vicinity of the walls and ground.

Once the calculations are made and the model adequately converges the post processing of the data files will allow the analysis of the pressure and velocity fields. The CFD head losses will be calculated following the Bernoulli equation (7) as can be found in any Fluid Mechanics book [10]:

$$E_1 = E_2 + h_p = \frac{V_1^2}{2g} + \frac{P_1}{\gamma} = \frac{V_2^2}{2g} + \frac{P_2}{\gamma} + h_p \rightarrow h_p = E_1 - E_2 \qquad (7)$$

where
E: energy (m)
V: flow velocity (m/s)
P: pressure (Pa)
g: gravity (m/s^2)
γ : specific weight (N/m^3)
h_p: head loss (m)

Quantities using a subscript “1” refer to a point in the centre of the tunnel at the entrance. A quantity with a subscript “2” refers to a point at the output.

4 Comparison: theoretical vs. CFD

Dozens of simulations have been done in the 500 m long tunnel domain. Although a condition of symmetry can be applied in the centre of the tunnel, thus allowing the use of half the number of meshing elements, the complete section geometry has been used in order to test the behaviour of the software in this 500 metres section.

Now there will be shown some of the results organized as per the mesh where they were calculated. Figure 4 shows the 4 meshes initially developed in order to accomplish the mesh independency study. Mesh 1 is a fine mesh composed only by tetrahedrons (“tetras”) and pyramids, using mainly tetras of 0.97 m edges. This creates approximately 400.000 elements in the mesh (400k) all over the tunnel domain. As approximately each element involves 1 Kb of RAM total amount of RAM needed in the calculation will be in the 400 Mb ranges, which is easily accessible for any nowadays computer.

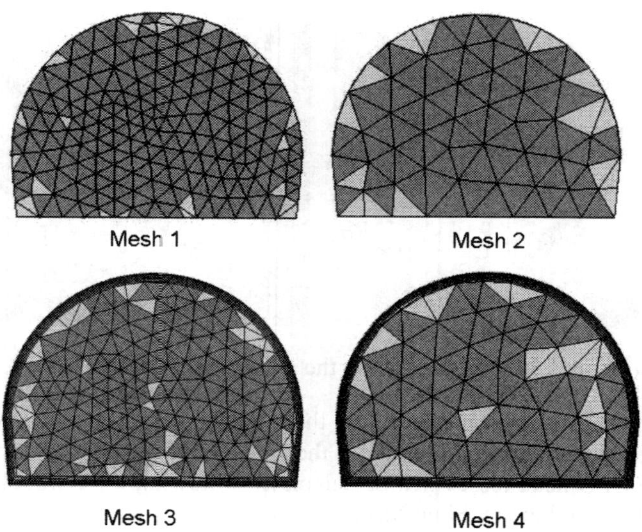

Figure 4: Different meshes used.

Mesh 2 has tetras of 1.2 meters of edge, thus obtaining less resolution, 225k elements. Mesh 3 adds to mesh 1 a layer of prisms around the tunnel walls and ground, creating 400k tetras and around 500k prisms. Memory and calculation times get increased. Mesh 4 adds prisms to mesh 2, obtaining 225k in tetras and 200k in prisms, lump sum of 425k elements. In all cases models obtained can be used in affordable time in computers with at least 1 Gb of RAM with no need to parallel processing.

Table 2: Mesh characteristics.

Mesh	1	2	3	4
Elements	477,793	253,328	918,338	426,224
Tetrahedral	409,826	226,697	397,425	225,932
Thickness prism layer (m)	N/A	N/A	0.292	0.292
Volume tetra (m^3)	0.106	0.17	0.109	0.192
Edge of tetra (m)	0.97	1.13	0.98	1.18

Each one of the meshes served to several calculations at several air velocities (between 0.2 and 8 m/s, ranges legally established in Spain by the underground works standards [12]) and all the roughness values included in Table 1 in case of walls, but maintaining null the roughness of the ground. This guide to several pair of data (CFD calculation, theoretical calculation as referred in point 2 of this paper) that can be compared in graphics as the ones in Figure 5 and 6. A regression line is fitted to each group of points. Following table 3 includes the regression line for each mesh (y=vertical coordinate=CFD calculation; x=horizontal line=Theoretical calculation).

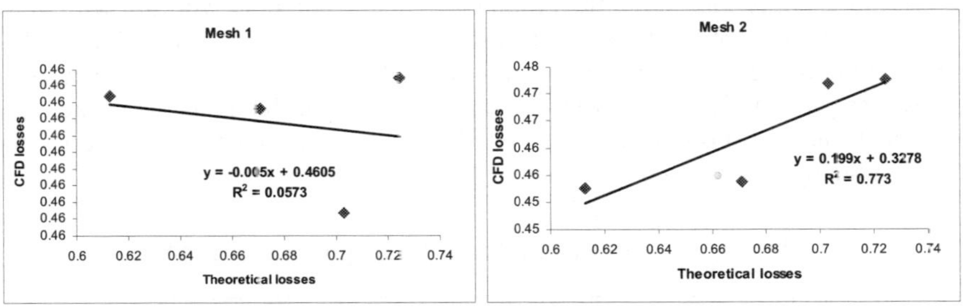

Figure 5: Comparisons 1 and 2.

The adjustment in case of mesh 1 is quite bad, with no correlation among several velocities and roughness. In case of mesh 2 results are better with correlation factors over 0.7 but with a line quite far from the ideal y=x fit.

Once prisms are included the results gets much better, with optimal regression factors over 0.99 in both mesh 3 and mesh 4. Seems clear that in this kind of simulations the use of prisms is a must. Mesh 3 and mesh 4 gives similar regression lines in $y=0.325 \cdot x$. This is, for any air velocity or wall roughness the losses obtained using CFD are the 32.5% of the losses obtained through the Darcy equation. This variation should come from the fact that the Darcy equation considers only one friction factor for both the walls and the ground of the tunnel, while CFD can take different friction factors and only walls friction factor has been considered.

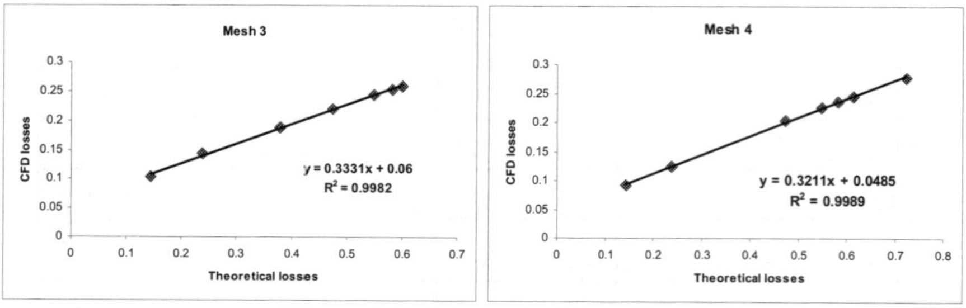

Figure 6: Comparisons 3 and 4.

Table 3: Error adjustment lines and regression factors.

Mesh	Regression equation	R^2
1	-0.005x+0.4605	0.0573
2	0.199x+0.3278	0.773
3	0.3331x+0.06	0.9982
4	0.3211x+0.0048	0.9989

If now there is considered the ground roughness, and only working with mesh 3, there will be obtained different adjustments as are shown in table 4. The regression line gets proximal to y=x, always with very good regression factors.

As the ground roughness grows so does the slope of the adjustment line. This can be explained through the fact that the head losses must grow as does the friction factor.

Table 4: Adjustment for different ground roughness.

Ground roughness	Adjustment line	R^2
Null	0.3211x+0.0048	0.9989
10 mm	0.5039x+0.0386	0.9993
Same as wall	0.6555x+0.0041	0.9991

5 Conclusions

Alter the calculations referred above first conclusion is that the calculation of pressure drops in tunnels requires tetrahedral meshes that include 12 tetras in the distance between the floor and the ceiling of the tunnel (what means in a 67 m^2 section tetras of 1.18 m of edge). The use of prisms in the neighbourhood of the walls is a must.

Using this two modelling advices there is obtained independency of the CFD results to the mesh used, which is mandatory when dealing with this sort of calculations. Hybrid meshes (by "hybrid" we mean concurrent use of tetras and prisms) of at least 400k elements give the same results as the meshes of 900k elements. The CFD calculated values differ from the theoretical calculations in a 68% value when considering no roughness of the ground. But the regression factor of the adjustment line of the comparison between both values (see figures 5 and 6) is almost 1, what lead us to say that calculations are right.

The roughness of the ground will affect calculations, as the CFD pressure drop vs. calculated pressure drop ratio will grow as the ground roughness do. If the same roughness is used in the ground and in the walls (which fits better with the concept of hydraulic diameter) a ratio of 0.65 is obtained. More simulations are being conducted in order to improve and explain this results.

Acknowledgements

We want to acknowledge the support from the Spanish Ministry of Science and Education that granted these researches through the project CTM2005-00187/TECNO, "Prediction models and prevention systems in the particle atmospheric contamination in an industrial environment".

References

[1] J. Toraño et al - A CFD Lagrangian particle model to analyze the dust dispersion problem in quarries blasts. Capítulo del libro «Computational methods in Multiphase Flow IV». 2007. ISBN: 9781-84564-079-8

[2] J. Toraño et al, "Contamination by particulated material in blasts: analysis, application and adaptation of the existent calculation formulas and software". Environmental Health Risk III, pp. 209-219, (2004)Sd

[3] J. Toraño et al,"Computational Fluid dynamic (CFD) use in the simulation of the death end ventilation in tunnel and galleries. Advances in Fluid Mechanics VI, pp 113-122.

[4] K.W. Moloney, I.S. Lowndes and G.K. Hargrave, "Analysis of flow patterns in drivages with auxiliary ventilation", Trans. Instn Min. Metall. 108, pp 17-26 (1999).

[5] S.A. Silvester, I.S. Lowndes and S.W. Kingman, "The ventilation of an underground crushing plant", Mining Technology (Trans. Inst. Min. Metall. A), Vol. 113, pp. 201-214 (2004)

[6] Achieving Energy Efficiency for Underground Mine Ventilation Systems. A.J. Basu, D. Datta. Web link. www.miningmiair.com/M%26I%2520-%2520MEMO-Sudbury%25202005.pdf, accessed 01/04/07.

[7] Meyer, CF. Determining the friction factors for underground colliery board and pillar workings. Safety in Mine Research Advisory Committee, Col 465, May, 1998, pp 1-84

[8] Ke, Ming-Tsun et al, Numerical simulation for optimizing the design of subway environmental control system. Building and Environment 37 (2002) 1139 – 1152. Elsevier.

[9] Gao, P.Z. et al. Large eddy simulations for studying tunnel smoke ventilation. Tunnelling and Underground Space Technology 19 (2004) 577–586
[10] Mechanics of Fluids 3rd. Merle C. Potter. David C. Wiggert. Ed. Brooks/Cole. (2001)
[11] Hacar, F. La rugosidad en túneles sin revestimiento en relación con la ventilación. Ingeotúneles Libro 3. Pp 455-491. ETSI Minas Madrid. (2000)
[12] Spanish Standard RGNBSM, ITC 05.0.01

Fragments of a buried urban past revealed through multi-layered voids hidden below the mosque of St. Daniel: the case of the underground museum in Tarsus

M. Cetin[1] & S. Doyduk[2]
[1]*Department of Architecture, Yeditepe University, Turkey*
[2]*Department of Architecture, Yildiz Technical University, Turkey*

Abstract

The historical process of urban-architectural layering in cities appears to accentuate the role of the intersections between various networks of urban circulation and public spaces from different eras. One of these spaces was recently unearthed in southern Turkey, where many civilizations have accumulated over the course of time. This underground building complex consists of cavities in addition to the remains of the foundations of a 16th century Ottoman bath, as well as the tomb of St. Daniel next to a Roman bridge vault that were recently excavated below 19th century Makam Mosque in Tarsus. The spatial formation here displays an extraordinarily complicated three dimensional stratification below the ground. This paper begins with historical research and analysis of space and continues with a design proposal to fuse all the religious, historical, geographical, architectural, spatial and material content into a single tectonic entity. Therefore, in summary, the paper addresses the issues of multi faceted design criteria regarding tangible and intangible aspects, and further discusses the issues such as; how such physical contexts enable multiple readings of history through spatial configuration, how geometrical grammar operates to narrate the history of urban stratification and how state-of-the-art architectural and engineering technology co-exist in historic contexts.

Keywords: urban archaeology, architecture, conservation, space, tomb of St. Daniel, museum, Tarsus, Turkey.

www.witpress.com, ISSN 1743-3509 (on-line)
doi:10.2495/US080141

1 Introduction

Tarsus is a town located in the southern part of Turkey, and has been a major settlement centre during the civilisations of Rome, Cilicia, Seljuk and the Ottoman empires [1–3]. Thus, it has witnessed a comprehensive urban stratification throughout the ages, which elevated the current altitude (in other words, ground level) of the city approximately 7 m above its original level during Roman times [4,5]. The archaeological excavations conducted in and around the 19th century Makam–i Daniel mosque revealed a complex spatial structure (Fig.1) of an underground spatial configuration (Fig.2) dating from periods such as the 1st, 7th, 13th, 16th, and 19th centuries A.D. Local authorities, and the conservation council demanded a genuine design solution developed to restore, conserve and to display these findings as they have special religious significance, particularly for Jewish and Muslim communities as well as visitors to the tomb of St. Daniel. Such a design should not only critically interpret and abstract the ongoing process of urban layering, but also contribute to such formation via its spatial and geometrical organisation. One of the major problems was in organising this space as a multi-religious cultural centre, while an associated problem was to unite sub-ground levels with ground floor facilities. Another problem was to construct a protective cover without obstructing the existing mosque building, which is of a significant local heritage.

Figure 1: Archaeological findings of the tomb of St. Daniel below the ground level of the 19th century Makam mosque in Tarsus.

Figure 2: Underground archaeological space to be converted to an exhibition hall.

Finally problems regarding the construction within the historic tissue and within the underground context, places the significance of the project onto an architectural version of intricate implant surgery.

2 Architectural intervention as an implant surgery to the archaeological tissue

The design proposal tackled this set of major problems; firstly, through various methods of archaeological analyses [6–9], and secondly, through a comprehensive survey of existing building and archaeological voids. The study further included a thorough review of precedents of such visionary projects. Therefore, the project consisted of both restoration work [10–13], and an architectural addition [14–17], which is also required to accommodate contemporary facilities for a museum of urban archaeology.

The project addresses the issue of utilization of this underground archaeological space as an intervention [18–24], the design concept of which manifests itself in the form of a museum of archaeology wrapping around a sacred core conserved in its authentic location. Hence, a unique spatial experience is intended whereby divine content and historical substance are intertwined into a ceremonial and mysterious hybrid of the multi-cultural physical context. Architectural design not only attempts to achieve a new and exciting sub-ground-level urban void, but also to equip a pedestrian node as an integral part of the urban living environment [25–29]. The architectural design theme is composed of a simple, neutral steel shell and wide glass screens as a protective structure (Fig.3) over the intricate labyrinth of interconnected sub-ground spaces. It also regulates the spatial experience around a spinal

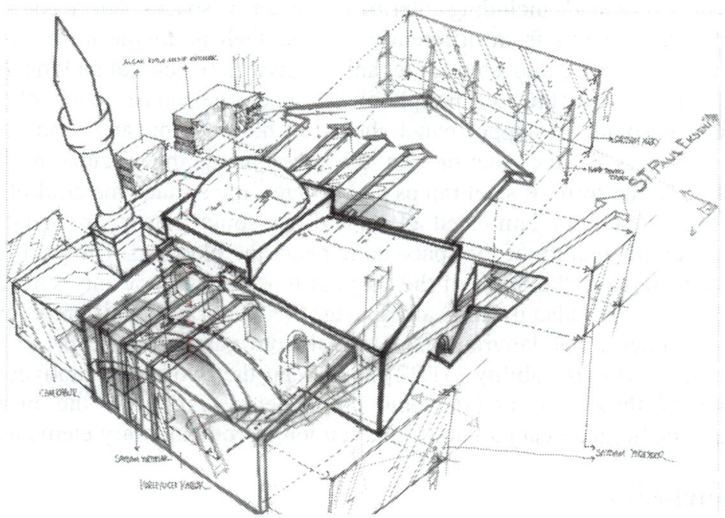

Figure 3: Schema illustrating the principles of the protective shell over the underground archaeological space.

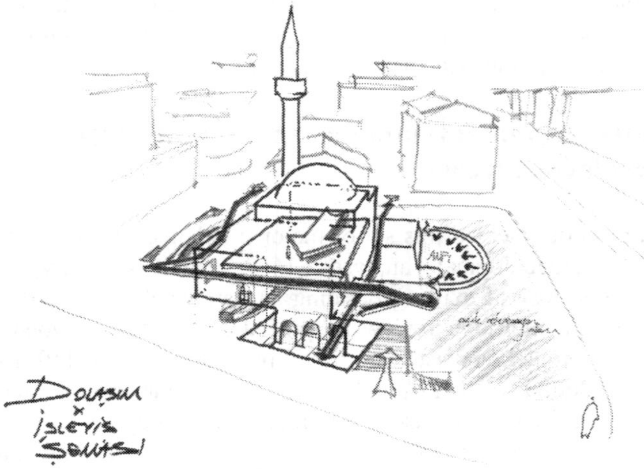

Figure 4: Circulation scheme of the proposed underground museum.

component, that is to say, an uninterrupted three dimensional circulation path (Fig.4) via steel and glass decks and bridges laid out so as to unveil the historical layering through human movement.

The design proposal re-interprets the underlying grammar [30] of the process of urban stratification through the geometry of the new addition. It develops a composition based on the geometrical superimposition in accordance with dominant urban axes and orientation of existing spatial configuration (Fig.5). This addition is basically a protective shell (Fig.6), uniting spaces both below and above the ground, including interior and exterior spaces that accommodate archaeological remains from different eras. The shell is double layer structure, accommodating a void allocated for technical services (electrical and mechanical facilities) in between the two peripheries. It is a steel construction clad with composite pre-oxidised copper panel sheets on the exterior, and with compact laminated panels on the interior. On the other hand, the circulation path is constructed in the form of steel ramps and bridges, the surface material of which ranges from wood, to laminated glass and, in some places, to metal mesh depending on the quality of the space underneath the platform. The path follows a route starting from the north of the original mosque that not only surrounds it from three sides, but also dynamically locates the visitor to the different levels in this three dimensional labyrinth-like spatial configuration. The principles of transparency and permeability [31,32] have been the two major motives in the formation of these two architectonic components (Fig.7) as the means of honouring the heritage via judicious intervention of contemporary elements.

3 Conclusion

The public demand for the utilisation of a masterpiece of cultural heritage consisting of architectural spaces and underground archaeological remains is

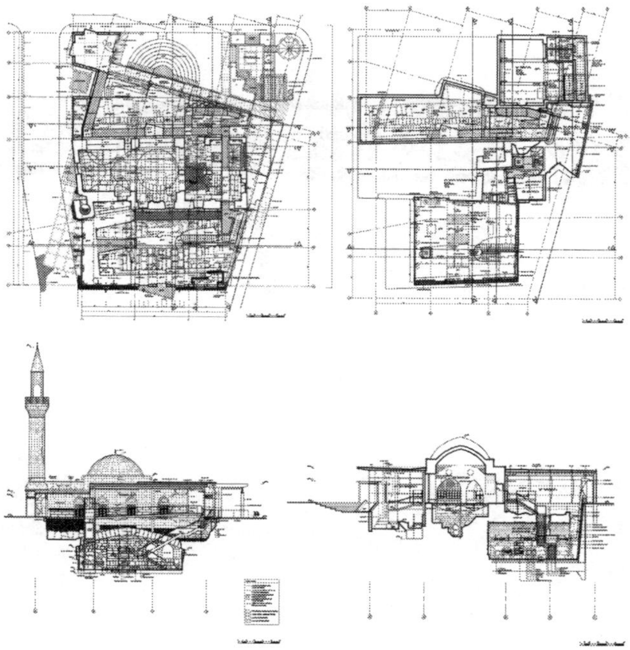

Figure 5: Plans and sections of the proposed museum building.

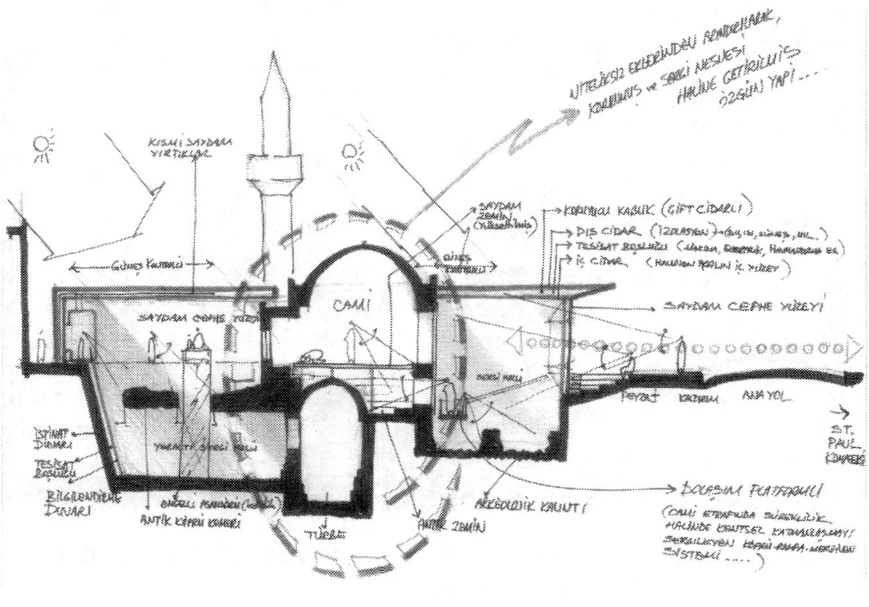

Figure 6: Section showing the 3D quality of the underground space.

www.witpress.com, ISSN 1743-3509 (on-line)

Figure 7: Sketch showing the transparent and permeable spatial quality of proposed underground museum.

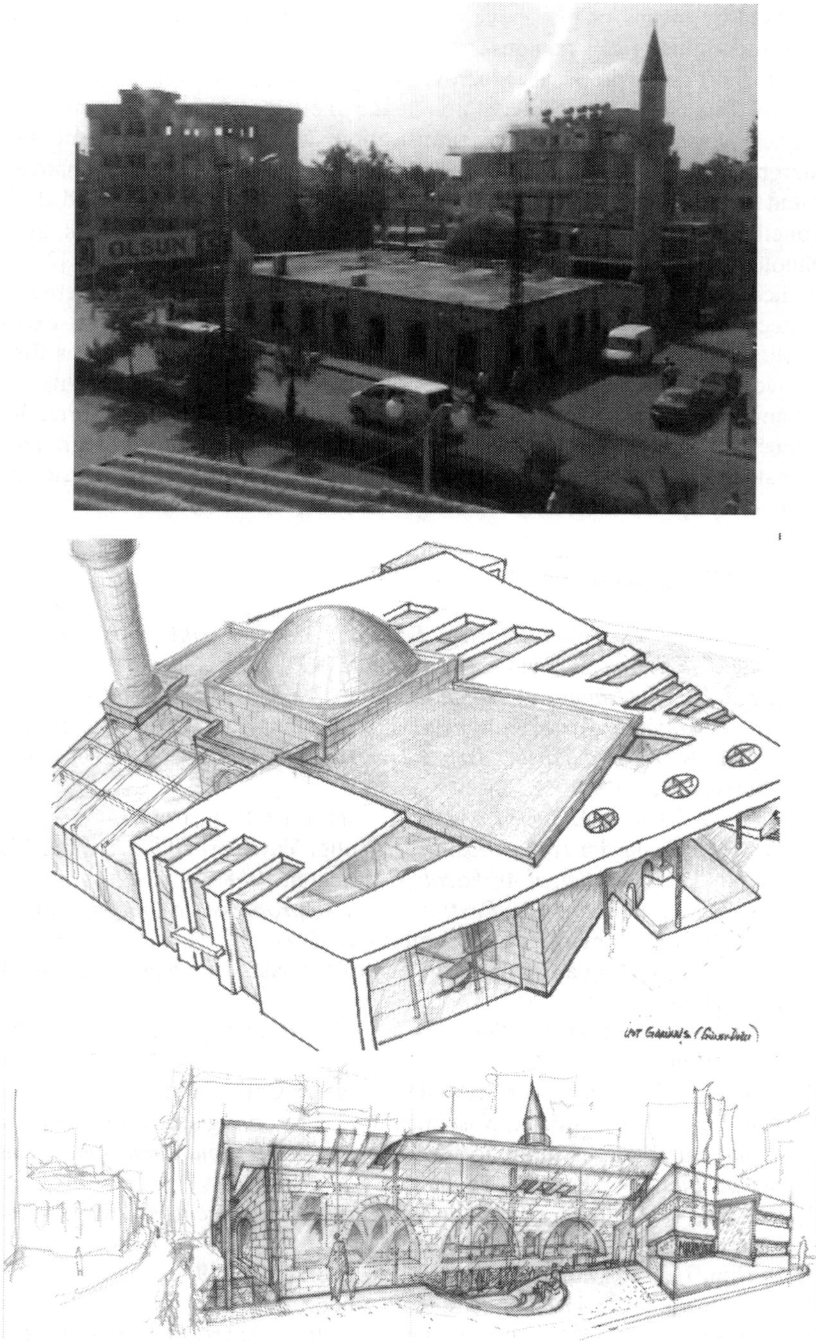

Figure 8: Existing outlook of the mosque and sketch of proposed museum.

addressed by means of the above discussed design project. In this framework, underground historical remains are integrated with current urban life [33] through the design proposal. Moreover, fragments of the urban history of the town of Tarsus are deciphered through the movement of visitors among the historical layers of the urban stratification within the three dimensional underground space which is intertwined with the architectural spaces above the ground around this mosque (Fig.8). The intended aesthetics of hybrid design, reconciling; old and new, tradition and innovation, conventional typology and technology, as well as spaces below and above layers of ground [34] are searched for within the intersection area of the disciplines of contemporary architectural-engineering design and heritage conservation. Thus, an existing complicated spatial asset that is located in a multi-cultural geography is further improved within a multi-disciplinary synthesis. Consequently, the historical continuity of the ongoing local urban stratification is sustained through an architectural design that expresses its multi-layered content through all its dimensions ranging from plannimetric spatial configuration and geometric shape grammar to façade treatment and articulation of architectonic massing.

References

[1] Akgunduz, A., *Arsiv Belgeleri Işıgında Tarsus Tarihi ve Eshab-ı Kehf* Tarsus Ticaret ve Sanayi Odası: Istanbul, 1993.
[2] Bilgili, A.S., *Osmanlı Doneminde Tarsus Sancagı ve Tarsus Turkmenleri; Sosyo-Ekonomik Tarih*, Kultur Bakanligi Yayinlari/2657: Ankara, 2001.
[3] Ciplak, M.N., *İçel Tarihi; Tarihi Turistik Zenginlikleriyle*, Guzel Sanatlar Matbaasi: Ankara, 1968.
[4] Erzen, A., *Tarsus Kilavuzu*, Maarif Matbaasi: Istanbul, 1943.
[5] Oz, H., *Bilinmeyen Tarsus*, Kultur Bakanlıgı Yayinlari 2038: Ankara, 1998.
[6] Clarke, D.L., *Analytical Archaeology*, London, 1968.
[7] Hodder, I., *Reading the Past; Current approaches to Interpretation in Archaeology*, Cambridge UP: Cambridge, 1986.
[8] Hyett, P., Building a future for the city of the past. *Architect's Journal*, **203**, p.25, 1996.
[9] Larkham, P.J., *Conservation and the City*, Routledge: London and New York, 1996.
[10] Boito, C., *On Restauratores*, Atelier Editorial: Cotia, 2003.
[11] Amorim, L., Loureiro, C., Nascimento, C., Preserving space; towards a new architectural conservation agenda. *Proc. of 5th International Space Syntax Symposium*, ITU, Istanbul, pp. 032/01–13, 2007.
[12] Pickard, R.D., *Conservation in the Built Environment*, Addison Wesley Longman Ltd.: London, 1996.
[13] Ruskin, J., *Seven Lamps of Architecture*, London, 1849.
[14] Schleifer, S., *Converted Spaces*, Koln, 2006.
[15] Worskett, R., 1984, "New Buildings in Historic Areas; The Missing Ethic", *Momentum*, V.25, s.29-154.
[16] Thiebaut, P., 2007, Old Buildings Looking for New Use, Stuttgart.

[17] Warren, J., Worthington, J., Taylor, S. (eds.). *Context; New Buildings in Historic Settings*, Architectural Press: Oxford, 1998.
[18] Bedard, J.F., *Cities of Artificial Excavation; Works of P. Eisenman 1978-88*, Rizzoli: Montreal, 1994.
[19] Borden, I. & Dunster, D. (eds.). *Architecture and the Sites of History; Interpretations of Buildings and Cities*, Butterworth: Oxford, 1995
[20] Byard, P.S., *Architecture of Additions*, W.W. Norton & Company: London, 1998.
[21] Cantacuzino, S., *Re-Architecture; Old Buildings, New Uses*, London, 1989.
[22] Cramer, J. & Breitling, S., *Architecture in Existing Fabric; Planning, Design, Building*, Birkhauser: Basel, Boston and Berlin, 2007.
[23] Eisenman, P., *Written Into the Void*, Yale UP: New Haven & London, 2007.
[24] Groat, L., Contextual compatibility in architecture. *Ethnoscapes,* ed. D. Canter, *et al*., Aldershot: Gover, 1988.
[25] Jessen, H. & Schneider, K., Conversions – the new normal. *In Detail: Building in Existing Fabric; Refurbishment, Extensions, New Design,* ed. C. Schittich, Birkhauser: Basel, Boston and Berlin, pp.11-21, 2003.
[26] Latham, D., *Creative re-use of buildings, Vol. I; principle & practice*, Shaftesbury, 2000.
[27] Mastropietro, M., *Restoration & Beyond; Architecture from Conservation to Conversion*, Milan, 1996.
[28] Powell, K., *Architecture Reborn; the Conversion and Reconstruction of Old Buildings*, London, 1999.
[29] Schittich, C., *In Detail: Building in Existing Fabric; Refurbishment, Extensions, New Design*, Birkhauser: Basel, Boston and Berlin, 2003.
[30] Çetin, M., "A formal grammar analysis of urban transformation", *Unpublished PhD. Thesis*, Sheffield University, Department of Architecture, 1999.
[31] Strike, J., *Architecture in Conservation*, Routledge: New York, 1994.
[32] Hoesli, B., Commentary on Colin Rowe. *Transparency,* ed. R. Slutzky, Stuttgart, 1968.
[33] Doyduk, S., Akkor, G., La participation de la stratification historique du Tarsus dans le tissue traditionnelle au mémoire public, *1st Euro-Mediterranean Regional Conference,* RehabiMed: Barcelona, 2007.
[34] Amorim, L. & Loureiro, C., On the spatial dimension of modern architecture as an object of conservation. *Proc. of International Seminar on the Management of the Shared Mediterranean Heritage, 5th Conference on the Mediterranean Heritage*, IRD: Alexandria, 2005.

Increase of stability of underground works

K. Weiglová[1] & P. P. Procházka[2]
[1]*VUT in Brno, Institute of Geomechanics, Brno, Czech Republic*
[2]*CTU in Prague, Civil Engineering, Prague, Czech Republic*

Abstract

In this paper the influence of additional structural precaution is studied for an exceptional tunnel facing in extraordinarily difficult conditions. In the case considered here artificial roofing is created by a system of horizontally drilled reinforced piles, which are prepared before drilling current purchase in the frame of tunnel engineering. They are positioned at three height levels above the expected vault (calotte) of the future tunnel, but in the model only one level is assumed (the others are built up for the sake of safety). The influence of the roofing has been verified on two independent models (with and without roofing), treated both experimentally and numerically.

Both physical scale models (experimental) and the numerical approach confirmed extremely high improvement of long-lasting stability of the underground work and settlement of the overburden. Using coupled modeling (experimental and numerical) enabled one to quantify the values of the necessary parameters. In numerical models the influence of pile roofing was measured in deformation and stress states of the rock mass. From the results obtained during numerical experiments with different types of roofing the deformation of the overburden was basically restricted. Depending on the material, the deformation dropped by as much as 95 percent in comparison to the non-reinforced case.

In the paper the approach consisting of couple modeling – experimental and numerical – will be described. The experiments are prepared on scale models in stands (basins with glazed front sides and the length of 2–6 m), where physically equivalent materials simulate the real situation. Based on similarity rules very good agreement with reality is attained.

Keywords: scale modeling, stability of underground works, tunnel face, coupled modeling, artificial roofing.

www.witpress.com, ISSN 1743-3509 (on-line)
doi:10.2495/US080151

1 Introduction

Coupled and comparative models consist of connection of results from mathematical and experimental models. In the experiments physically equivalent materials are used, which simulate real states in reality. In order to get the real situation in the scale model, similarity rules have to be fulfilled, [1, 2]. In this way the experimental model becomes real. The structure of tunnel behavior in extreme conditions is studied in this paper. The tunnel is built up in the city of Brno, Czech Republic, and the conditions which should be taken into consideration require involvement of predisposed cracks, which appear in the neighborhood of the tunnel lining. In the mathematical model the eigenparameter technique is exploited [3]. This approach is generalized starting with nonlinear behavior of the rock, which is described by Mises hypothesis [4]. Previously [5], such a combination of eigenparameters and starting plasticity state was applied. For details see [6], where coupled modeling based on mutual affection of scale experimental models and numerical models based on the generalized eigenparameter technique is described.

Here the influence of predisposed cracks on tunnel stability is studied. In this sense instead of eigenparameters the coefficient of internal friction φ and the cohesion (shear strength) c are considered as free parameters in the coupled modeling. The basic idea of the former approach is still considered, i.e. influence matrices are first created for the fully elastic case and a linear hull of plastic, eigenparameter and elastic stages are considered in such a way that the measured values in experimental models and that of numerical models at selected points are as close as possible. For this reason, unit internal parameters (the coefficient of internal friction) and shear strength are introduced. As the influences of the latter parameters are solved in elastic (linear) medium, influence matrices are easy to construct. The auxiliary functional of variances of differences in stresses or displacements is formulated and minimized for unknown values of the design parameters φ and c. Once they are determined the critical situation can be assessed, which covers visible damage of the structure, strain observation and possible limitation, and, eventually, failure of the structure.

2 Description of the situation on site

In the city of Brno, Czech Republic, Tunnels Dobrovského (Street) of the Great City Circuit will be an important component of the communication system of the city as well as of the road network of the Czech Republic and of an international network. The tunnels are driven in the length of about 1000 m. The tunnels are engineered in parallel with the axial distance of about 80 m. The thickest complex with the largest area of Neogene sediments in Brno and surroundings is chalky clay, often with a perceptible disintegration, so-called fissured texture fine grained soil with very high plasticity. Volume instability is expected in this are of construction. In Fig. 1 the longitudinal cross-section is seen with various layers of geological positions.

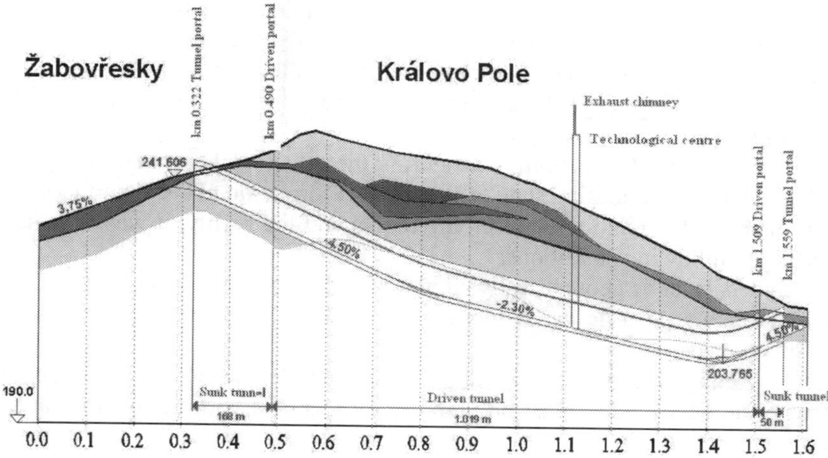

Figure 1: Longitudinal cross-section of the tunnel.

Calcareous clay containing 10–30% of carbonate minerals are the prevailing minerals on site. An admixture of classic grains (quartz, muscovite, feldspars) is contained in the rock, as well as the calcareous to siliceous shells of microfossils. Pyrite and organic substances, clayey minerals in the amount of 50–60% and the presence of smectites of the illite-chlorite type also state the essential properties of the rock.

Unlayered or inconspicuously layered sediment sherd-like break-off to shelly zones with the fissured texture surfaces of discontinuities of different directions are observed with piece-like slopes disintegration in zones of weathering pelitic microstructure granulation of 0.002–0.0005 mm. From the point of view of contraction and swelling the Brno clay is one of the most problematic soils. The reason is the presence of smectites Neogene clay Brno-Královo Pole: high values of the plasticity index, high values of the activity index (2.2) and the cofactor of volume swelling B = 7%, sensitivity St = 1.3 (at rigid consistency) – which results in development of swelling pressures.

In Fig. 2 results from triaxial tests are depicted showing basic mechanical properties of the material.

3 Experimental

The scale models are prepared to deliver data for accurate numerical modeling of problems required in geotechnical engineering. Back analysis or coupled modeling is one of the possible approaches to identify deformation of tunnel lining, tunnel surrounding rock, crack occurrence and finally, identification of failure of the structure. In our case of extremely exacting conditions (shallow foundation of the tunnels, clayey soil, loose rock, etc.) extreme attention has to be drawn to other certain special items:

- Establishing the stability of the provisional rock pillar (middle pillar) between the galleries when driving the tunnel by vertical articulation of the stope including the determination of the optimum length of the attack.
- Examination of the rate of pushing up the footwall into the floor up to the failure of the provisional rock pillar.
- Whole tunnel profile (section), considering the effect of the lining, the system was loaded up to the limiting state of failure with formation of shear surfaces.

To maintain the relation between the model and the reality, it was necessary to fulfill the following conditions:
- The models must be geometrically similar to reality, see the following text
- Actions taking place in the model and in the work must belong to the same class of actions
- The beginning and marginal conditions in the model, which are expressed in the dimensionless model, must be numerically identical with the dimensionless conditions in the work.

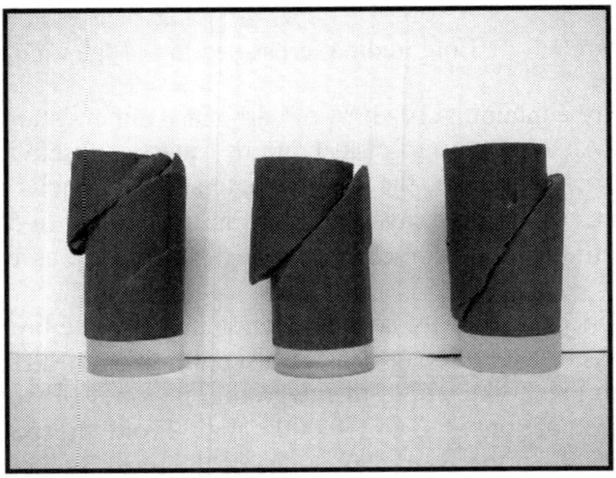

Figure 2: Samples of Brno clay after the test in the triaxial apparatus.

The dimensionless arguments of the same nature must be numerically identical in the model and in the work.

Models of identical materials are built up on the basis of dimensional analysis and in accord with the theory of similarity, which make them possible to be feasible to the preliminary qualitative and theoretical analysis of problems and, at the same time, to the selection of the system determining dimensionless magnitudes.

In physical models several similarities have to be obeyed. One of the most important is geometrical similarity; it is proportionality of dimensions and identity of angles between model and modeled object in the whole range of the model. Altogether, possibility and technique of bringing of forces, time factor and other aspects, technical possibilities, assign linear scale of the model are objective of the modeled geotechnical problem. It is the aspect ratio, which plays essential role, in which length dimensions of the model are reduced against the reality.

If $1/\alpha_l$ is the length scale, i.e. the ratio of lengths on model and reality and a is the ratio of bulk densities

$$a = \rho_{model}/\rho_{reality}, \tag{1}$$

then it can be defined for

forces $\quad P_{model} = a.\ (1/\alpha_l)^3.\ P_{reality}$

stresses $\quad \sigma_{model} = a.\ (1/\alpha_l).\ \sigma_{reality}$

deformations $\quad \varepsilon_{model} = a.\ (1/\alpha_l).\ \varepsilon_{reality}.$

To simulate processes taking place in the rock material in the most perfect way, rock environment is replaced in the model from equivalent materials by their determinate physical and mechanical properties according to the model laws. The scale of model agrees with the rock properties and respects the character of failures simulating those in rock material. The models are constructed from mixture of various, mostly easy available materials: On the basis of an extensive set of laboratory experiments (ballotine, bentonite, glycerine, sand, graphite and ferrosilicium) physically equivalent materials were selected as:

- sand with an admixture of fat A00 (99.5% + 0.5%)
- ballotine with an admixture of fat A00 (99% + 1%)
- ferrosilicium with an admixture of fat A00 (99.5% + 0.5%)

The scale of the model is chosen according to the smallest part of the structure (in our case the thickness of the outfit), which must be feasible.

In Fig. 3 face view of the stand with the structure illustrate the model geometry. The reinforced basin (the stand) with glazed front wall shows possibilities of observation of changes in models (or, equivalently, in reality) not only from the point of view of quantity but also from the standpoint of quality, as time dependent processes can be qualified.

In Fig. 4 detailed view of artificial roofing and future tunnel are depicted. The experimental models in stands simulate the real situation on a scale 1 : 20. In the upper part of the picture the horizontal piles are seen, which are created from rebar reinforced concrete. They can be in 3D model considered as beams with given bending stiffness.

In Fig. 5 detailed view of the tunnel in the scale model (the stand) is shown with measurement equipments. Dilatometers are positioned inside of the tunnel hole and on the face of the lining also tensometer gauges are stuck.

4 Numerical approach

The aim of this paragraph is to show some possible relations involving stresses (or strains) and eigenstresses (or eigenstrains) in composite structures. Consider the domain $\Omega \in R^3$ with boundary Γ, describing the shape of the body under study. Inside this body n subdomains are situated $\Omega_1, \Omega_2, ..., \Omega_n$. The boundary Γ is split as follows: $\Gamma \equiv \Gamma_u \cup \Gamma_p, \Gamma_u \cap \Gamma_p = 0$.

On Γ_u the displacements $u_i = \overline{u}_i, i = 1,2,3$ are prescribed, while on Γ_p the tractions $p_i = \overline{p}_i$, $i = 1,2,3$ are given. The primed quantities are given.

Figure 3: View of the stand.

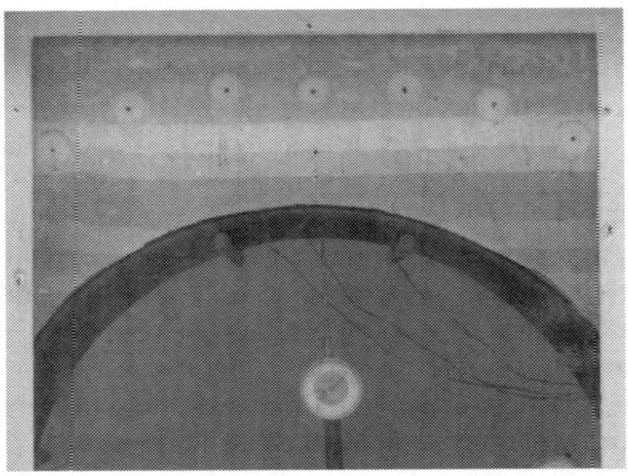

Figure 4: Detailed view of the stand with artificial roofing.

Let us consider that there is negligible plastic behavior in lining (being mostly compressed). In this case Hook's law with material stiffness L holds

$$\sigma_{ij} = L_{ijkl}\varepsilon_{kl} + \lambda_{ij}\,, \quad \lambda_{ij} = -L_{ijkl}\mu_{kl} \quad \text{in } \Omega \tag{2}$$

The real displacements $\boldsymbol{u}$, strains $\boldsymbol{\varepsilon}$, and stresses $\boldsymbol{\sigma}$ are unknown in the latter relation, so are the eigenstrains $\boldsymbol{\mu}$.

The symmetric polarization stress tensor $\boldsymbol{\tau}$ is introduced as:

$$\sigma_{ij} = L_{ijkl}^{\text{lining}}\varepsilon_{kl} + \tau_{ij} \tag{3}$$

Since both $\boldsymbol{\sigma}$ is statically admissible, it holds:

$$\frac{\partial \sigma_{ij}}{\partial x_j} = \frac{\partial (L_{ijkl}^{\text{lining}}\varepsilon_{kl} + \tau_{ij})}{\partial x_j} = 0 \quad \text{in } \Omega \tag{4}$$

$$\tau_{ij} - [L]_{ijkl}\varepsilon_{kl} - \lambda_{ij} = 0 \text{ in } \Omega \tag{5}$$

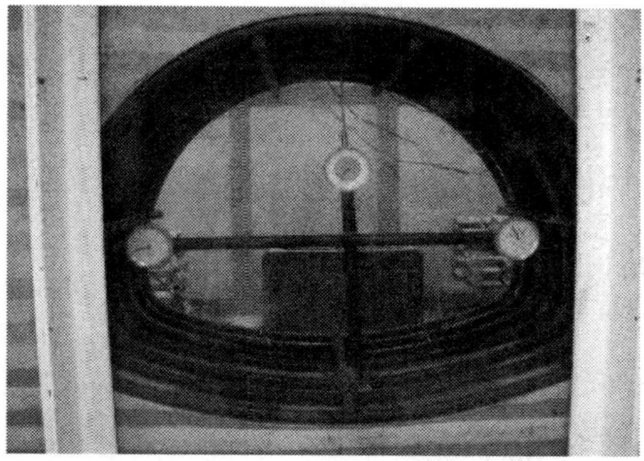

Figure 5: Detail of the tunnel with reinforcement and measuring equipments.

and

$$[L]_{ijkl} = L_{ijkl} - L_{ijkl}^{\text{lining}}.$$

Note that the relation (3) can be written in more details as:

$$\sigma_{ij} = 2G^{\text{lining}}\varepsilon_{ij} + \delta_{ij}\frac{2G^{\text{lining}}\nu^{\text{lining}}}{1-2\nu^{\text{lining}}}\varepsilon_{kk} + \tau_{ij} = 2G^{\text{lining}}\varepsilon_{ij} + \lambda^{\text{lining}}\delta_{ij}\varepsilon_{kk} + \tau_{ij} \quad (6)$$

and Kronecker's delta $\delta_{ij} = 1$ for $i = j$ and it is equal to zero otherwise. Material constants G and λ are Lame's constants (G is also referred to as the shear modulus), ν is the Poisson's number.

Applying the Green theorem to (4) after multiplying each i -th equation by φ_i, which is not identically zero in Ω and integrating over the domain yields

$$-\int_{\Omega}(L_{ijkl}^{\text{lining}}\varepsilon_{kl} + \tau_{ij})\frac{\partial\varphi_i}{\partial x_j}\mathrm{d}\Omega + \int_{\Gamma}(L_{ijkl}^{\text{lining}}\varepsilon_{kl} + \tau_{ij})\varphi_i n_j \mathrm{d}\Gamma = 0 \quad (7)$$

After decoding the latter equations in the sense of finite element method it is seen that the nonlinear behavior in the lining is suppressed and only rock has to be involved into plastic computation. Hence, using the back analysis (or coupled modeling) of experimental and numerical mutual influence, identification of eigenparameters are obtained for the rock without pile reinforcement. Using the material behavior from this state and extend it to the reinforced case, necessary length of the piles can be computed from the mathematical model. It is worth noting that parametric study on this problem of the necessary length of horizontal pile beam can hardly be conducted in experiment in the whole extent. On the other hand, numerical study is very cheap, not only from the standpoint of computation, but also because of very large extent of simulations of material and structural combinations.

5 Example

The numerical problem starts with the solution of plastic stage obeying generalized Mohr-Coulomb hypothesis with the following material parameters:

Modulus of elasticity $E = 1000$ MPa
Plastic $E = 800$ MPa
Residual $E = 500$ MPa
Poisson's ratio $\nu = 0.25$
Plastic equals residual $\nu = 0.46$
Shear strength $C = 0.08$ MPa
Plastic $C = 0.06$ MPa
Residual $C = 0.02$ MPa
tan $\varphi = 24^0$
Plastic tan $\varphi = 30^0$
Residual tan $\varphi = 10^0$
Volume weight $\rho = 2100$ kg/m^3.

Using the above said parameters describing material properties, which correspond with the class R_3 of the rock according to geological standards. The tunnel lining is made from concrete and no plastic behavior is assumed. After computing the plastic state with the above parameters, initial stage is created and the eigenstrains can be introduced.

The measured values were also vertical deflections on the contact of lining and rock. The values were taken from a scale model built up in stands. There were nine measurement points along the lining; the values at symmetric points were averaged.

In Fig. 6a) hypsography of vertical stresses in the domain without reinforcement is depicted and in Fig. 6b) hypsography of vertical stresses in final stage after optimization of the piles is displayed. Difference between these two pictures is not as much distinct as supposed.

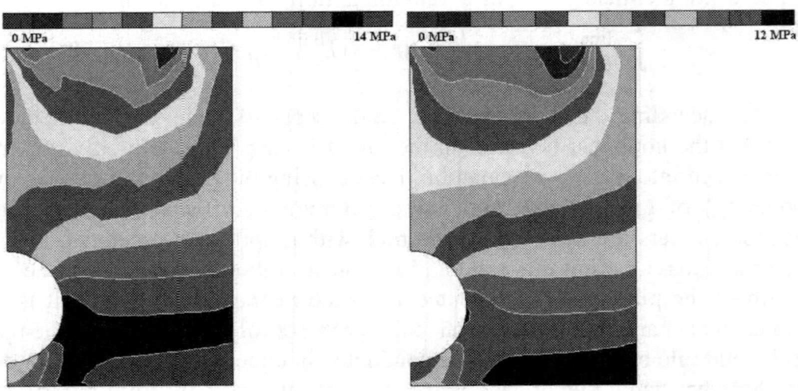

a) b)

Figure 6: Vertical stresses in a) plastic stage and b) after reinforcement.

In Fig. 7 relation between the length of pile beams and the length of purchase during tunneling is shown. The pile beams are designed according to classical beam's theory.

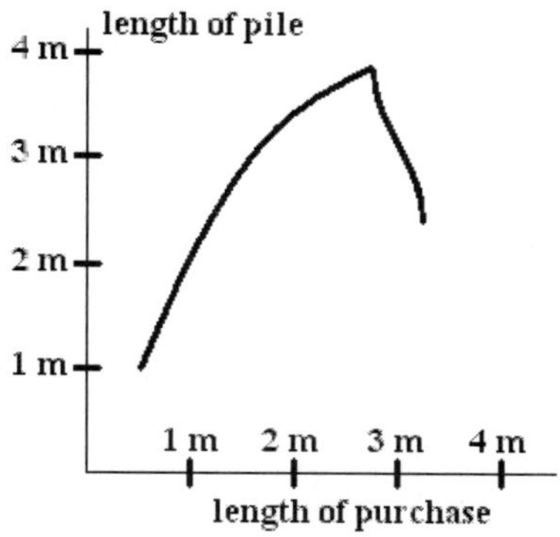

Figure 7: Length of pile necessary for stable engineering.

6 Conclusions

In this paper coupled modeling helps to assess tunnel face stability from the viewpoint of auxiliary reinforcement above the future tunnel (roofing). Scale models and numerical treatment lead to very close view into the situation on future site. The model, of course, verified and confirmed the right decision to build up roofing from the concrete reinforced piles. But not only this: they enabled us to determine necessary length of piles (considered as a beam in the 3D finite element model) to ensure the stability of engineering during tunneling. This necessity is induced by the fact that the reinforcing piles can be set up only during the process of construction of tunnels.

Acknowledgement

Work on this project has been financially supported by GACR, project No. 103/06/1124. Sponsorship of CIDEAS is also acknowledged.

References

[1] Kožešník, J. (1983). *Theory of similarity and modelling.* Academia, Prague, 1983.

[2] Head, K. H. (1992). *Manual of soil laboratory testing.* Wiley, New York, 1992.

[3] Dvorak, G.J. (1992). Transformation field analysis of inelastic composite materials. *Proc. R. Soc. London A 437,* 1992, 311–327.

[4] Duvant, J. and Lions, J.P. (1972). *Variational inequalities in mechanics.* DUNOD, Paris

[5] Procházka, P. and Weiglová, K. (2002). Coupled modelling using DSC and TFA models. *Proceedings of the ICTACEM,* Kharagpur, 2002, paper No. 105.

[6] Procházka, P. and Trčková, J. (to appear in *Solids and Structures*). Assessment and Control of Tunnel Structures based on coupled modelling.

Underground spaces and indoor comfort: the case of "Sassi di Matera"

A. Guida[1], A. Pagliuca[2] & G. Rospi[3]
[1]*Department of Architecture, Planning and Infrastructures for Transport, School of Engineering, University of Basilicata, Italy*
[2]*DAU, School of Engineering, Polytechnic University of Bari, Italy*
[3]*School of Engineering – Università Politecnica delle Marche, Italy*

Abstract

Underground architecture, like the "Sassi" in Matera, makes people imagine life in the past and reconsider present life in these places "without time" and how to "live" them, again like in the "past". This scientific research focuses on the quality of indoor living in this particular type of architecture.

Today indoor comfort is essential, because 85-90% of time is spent in closed spaces (rooms, work environment, etc.). Closed air, can be for people, more polluting and harmful than open air, as, inside we also have other harmful agents (germs, gas, dust and so on) whose danger is often undervalued.

Today it is very important while projecting to consider first of all the quality of indoor air and comfort.

The aim of this research is to analyse the thermal and hygrometer performance of underground spaces and to show how traditional Mediterranean masonry offers, also today, a high level of indoor comfort.

We did this with tests "in situ" monitoring indoor comfort (as low requests UNI EN ISO 7730 1997 and UNI EN ISO 7726 2002, by directive CEE n° 106/89).

The experiment was verified through the experimental applications in the restoration of two urban buildings in ancient "Sassi" in Matera, reconverted into hotels, the "Locanda di S. Martino" and "Hotel S. Angelo", that demonstrates as said before and how these building are suitable with the performance requirement that is required from residence environments (European Directive 2002/91/CE and Italian low D.lgs 192/05 – 311/06) even if they are built with traditional technologies.

Keywords: comfort indoor, air quality, underground space.

WIT Transactions on the Built Environment, Vol 102, © 2008 WIT Press
www.witpress.com, ISSN 1743-3509 (on-line)
doi:10.2495/US080161

1 Introduction

The action follows not simply the existing environment, but it turns producing a stratification of interventions based on management that are not always harmonious in the space. This process of transformation is an irreversible step, an action that is intended to permanently change the structure of a place changing also significantly the shape.

The instrument of this anthropization process of space is the stone which, with its unique technical characteristics and morphological-formal, becomes the obvious sign of this process.

The use of stone in architecture, in fact, has produced numerous applications, complex and highly heterogeneous between them from: the structures dug into the rock to built-up structures; initially they conceived as a natural continuation of the first, reaching then, their own formal autonomy and typological realizing the basic cell constructive called "lamia" or "lamione". And this is the process that has overseen the birth and development of the ancient districts "Sassi di Matera", in which the relationship between structures and built dug reaches its highest expression and formalization morphological.

The research included in a broader study, still ongoing is aimed to define the methodological and operational aspects of the recovery of these districts; part-type from morphological of this specific architectural context and aims to assess the possibility of obtaining performance requirements meet the quality standards today required by residence in building even in buildings with technical and constructive traditional features.

Through tests "in situ" aimed at monitoring of comfort "indoors" (as defined by UNI EN ISO 7730 1997 and UNI EN ISO 7726 2002, the EEC Directive N. 106/89), the study aims to highlight methods in which these spaces can guarantee good comfort conditions and considerable quality performance.

2 Indoor comfort

The term "indoor environment" is used to indicate all those confined environment of life and work that includes housing, offices, premises for recreational and/or social where people spend most of their life. Indeed, on average, the population spends more than 70% of their time in these environments undergoing, in fact, a prolonged contact with potential pollutant sources contained therein.

So the assessment of indoor comfort becomes a prerequisite for global comfort. This condition can be reached considering the air-quality comfort and thermo-hygrometric comfort.

The air-quality comfort mainly concerns the indoor concentrations of a variety of substances chemical predominantly nature: among these the most important ones are carbon dioxide, carbon monoxide, sulphur dioxide, radon, formaldehyde and all those volatile organic compounds that have serious consequences on the health of people.

It follows that to reach air-quality comfort we must maintain the values of concentration of these pollutants less than determinate minimum levels set by rule. This is possible mainly through a constant replacement of air which is dimensioned depending on the type of environments, their destinations use and the average capacity of people.

The thermo-hygrometric comfort, instead, is a function of a number of parameters: environmental (temperature, relative humidity, wind speed, solar radiation and atmospheric pressure), linked to physical activities (energy metabolic M), related to clothing worn (thermal clothing Icl) and the percentage of metabolic energy used for carrying out the physical activities (performance mechanical – value).

The current European regulations about thermo-hygrometrical performance classify indoor environments in three different types: environments "moderate" environments "severe hot" and environments "severe cold", each of which must comply with different levels of comfort.

The research focuses on thermal environments "moderate", namely those where the indoor conditions remain almost equal without considerable heat exchanges located between subject and environment that have significant effects on the overall heat balance (i.e. homes, offices, businesses, etc.).

They are characterized by thermal parameters variables within a limited range (as determined by the standard ANSI / ASHRAE 55-1992 - Thermal environmental conditions for human occupancy and UNI EN ISO 7730/1997 - Moderate thermal environments. Determination of PMV and PPD indices and specifications of condition of thermal comfort). They are:

- Air temperature (Ta) between 10 and 35°C;
- Temperature radiant (Tr) between 10 and 40°C;
- Relative Humidity (Ur) of between 30 and 70%;
- Air speed (Va) of between 0 and 1.5 m/s.

In environments "moderates" must assess the deviation of actual conditions than those comfort heat-humidity (temperature varies between 18-26°, relative humidity ranging between 50-60% and speed of air less than 1 m/s). The methodology used for assessing indoor comfort is out experimentally measurements using in-situ monitoring parameters microclimatic interior and surface temperatures.

3 The study case: underground spaces in "Sassi di Matera"

The site is a part of the architectural complex particularly significant, already a heritage of UNESCO since 1993: the ancient "Rioni Sassi di Matera", they are at an altitude of about 401m above sea level and are characterized by an aggregation of cells elementarily-called " the neighbourhood" that open towards the south-west.

The configuration of the complex morphology of the "Rioni Sassi" is that they can fully exploit the micro-climatic conditions of the site, characterized by hot and humid summers and mild winters.

The hypogea houses are arranged to horseshoe around a central atrium, (the "neighbourhood"), with exposure to the south; this exhibition allows a greater depth for the central caves, because they receive more sun and can more prolong inside the bench rock in which they are dug. The same homes are sloping inwards and so allow sunlight to penetrate in winter to the end, to heat and make healthy even the most hidden parts, while in summer, when the sun is highest, the most internal remain cooler.

In the site it is monitored three different types of hypogeous (with similar size and characteristics) of two hotel, the "Locanda di San Martino" and the "Hotel S. Angelo", different but similar at the same time in the forms and architectural peculiarities. They made it possible to compare environments with the same technological characteristics. In particular were considered a hypogeum not restored and a hypogeum restored, both in the "Locanda S. Martino" and a hypogeum restored and arrangements for use in "Hotel St. Angelo".

The monitoring was performed in the period from 10/04/2007 to 19/06/2007, a period characterised by a strong thermal excursion daily.

The instrumentation used was an environmental monitoring station with "multiacquisitore" LSI BABUC/A, equipped with 5 microclimatic probes:

- a probe globothermometric, for measuring the average temperature radiant;
- a probe psicrometric, for measuring the temperature of dry, humid temperature, temperature and dew on Relative Humidity;
- a probe anemometric portable hot– wire, for measuring the speed of the air;
- two thermometric probes (PT100), for– the measurement of surface temperature.

The measurements were carried out using fixed locations within the hypogea, placing the unit at a height of about 1.70 meters above the ground (average human height) and in the middle of the space.

In the hypogea restored and arrangements for use measurements were made with air conditioning off and considering two different conditions of operation of local: with plant recirculation of environment on and with plant recirculation of environment off.

For each local monitored were detected 7 signals with a frequency acquisition equal to 15 minutes, corresponding to the values of:

- Temperature dry-bulb air (Ta);
- Temperature wet-bulb to forced ventilation (Tw): from Ta Tw and the unit automatically gives the dew point temperature (Te) and relative humidity (Ur);
- Air speed (Va);
- Wet-bulb temperature and natural ventilation (Tnw);
- Temperature globo-thermometer of Vernon (Tg), from which the unit automatically derives the average temperature radiant (Tr);
- Surface temperature inside of the wall;
- Surface temperature of the external wall.

3.1 The behavior energy of not restored underground space

For architectures hypogea not restored means all those architectures that are in total state of disuse for many years.

And it is precisely the situation of total abandonment that often leads such architectures to situations of high relative humidity inside causing the formation of lichens and bacteria on the walls.

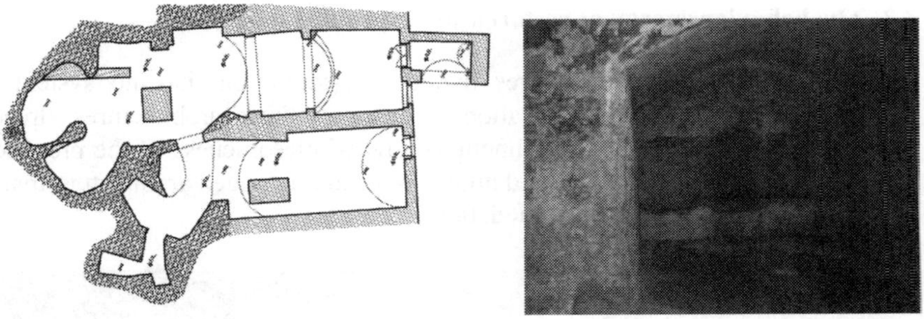

Figure 1: Not restored underground space.

The main characteristic of this type of environment is having a constant trend of microclimatic conditions during the indoor campaign measurements. Explicated in detail the individual values monitored we can say that the evolution of internal temperature had very small fluctuations (13-15°C), the average temperature radiant has remained constant during the measurement period (16°C), as well as the Relative Humidity (95%); as regards the speed and change of air, these values were next to zero.

Chart A refers to the entire period of monitoring.

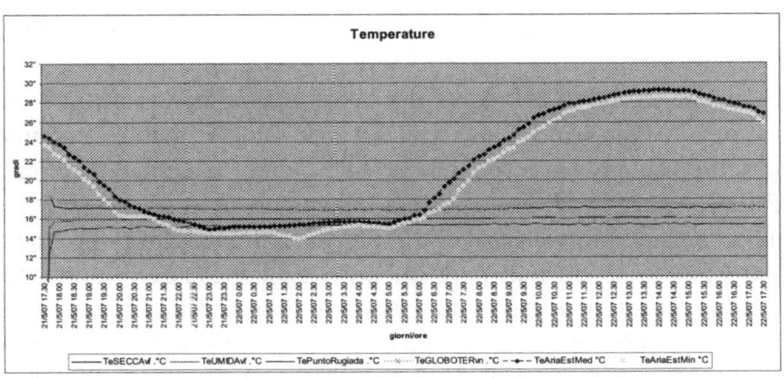

Chart A - Temperature of not restored underground space

As you can see from the charts, the slightest deviation of the values of dry-bulb temperature compared to the wet-bulb temperature and the dew-point temperature is the main cause high relative humidity indoor, allowing thus to lichens and bacteria to attacking the internal surfaces of the walls. In fact most of moisture present on masonry surfaces is caused to the continuous absorption of high humidity in the air a contact with the walls and a tiny amount of the migration of water present in rock.

Monitoring of these environments allows us to understand the nature of moisture present on surfaces walls of these environments, and we can say that in such environments humidity caused to rising capillary infiltration of water through the walls is virtually absent.

3.2 The behavior energy of restored underground spaces

Let us now consider architectures hypogea restored, but has no system of operation and use. This classification includes all those architectures already recovered from the aspect static-functional and wholesomeness of the premises. The values monitored, as described above, refer to a hypogea architecture that is part of a complex used as accommodation.

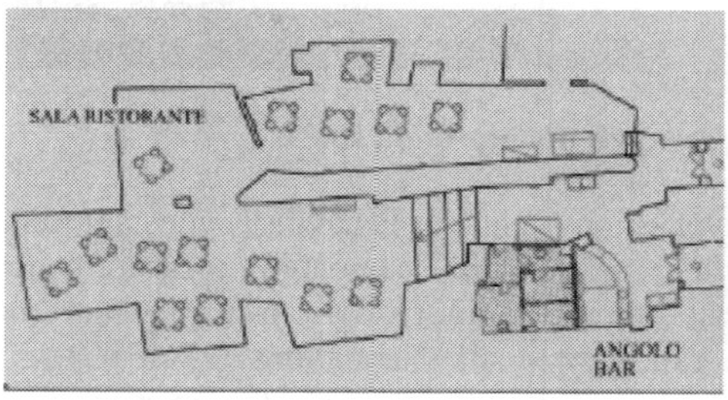

Figure 2: Restored underground space.

Again, the main characteristic of this type of architecture was to have microclimatic conditions indoor constant, even if with different values than the type not restored.

In fact the values of measured of the internal temperature air had increased by an average of 2°C compared to the type not redeveloped, stable at around 15°C, on the contrary the values of average temperature radiant remained identical (16°C).

The Relative Humidity, however, had a sudden drop of 20%, by reference to values more acceptable (about 65%), but still high respect on the standards. With regard to the speed and parts of air, again measured values are close to zero. The Chart B refers to the entire period of monitoring.
temperature, wet-bulb temperature and dew-point temperature, in the hypogeoum not restored is almost nothing. Indeed, this difference has contributed to lowering relative humidity environment.

This analysis has highlighted the importance of indoor microclimate control as a prerequisite to design a proper (and effective!) intervention rehabilitation thermo-hygrotermic of the “Sassi of Matera”.

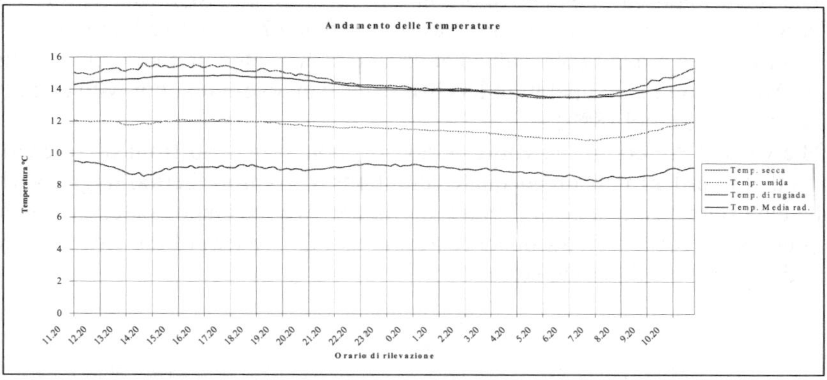

Chart B - Temperature of restored underground space.

Controlling indoor microclimate it will get comfort levels close to the minimum required by regulations of reference (indoor temperature of 18-20°C and relative humidity 50%).

3.3 The behavior energy of a rate of operation and use underground spaces

It is here described the behaviour thermo-hygrometric of a hypogea restored and operating system for many years.

In this configuration it was possible to make measurements with the planting of recycling and treatment of the environment on (UTA - air handling unit), which change air without conditioning. In this way the air blown into the air pushes vitiated air outwards in a natural way through the “light above” (small window above the front door), given appropriate open.

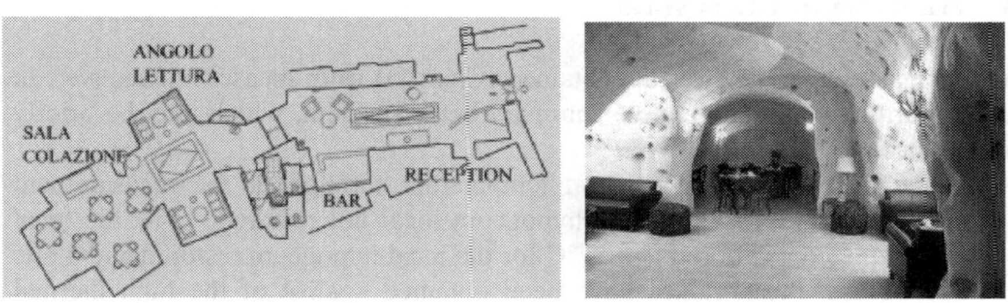

Figure 3: Rate of operation and use underground spaces.

The measurements made in this configuration made it possible to check the impact of the loop on the parameters of indoor comfort. It was also possible to verify the behaviour of environments hypogea in two different configurations: with recycling and fan system on and with recycling and fan system shut down.

The most important content found is that in those environments setting the UTA so as to change 5 volumes/day has obtained an drop relative humidity who rose from values around 65% valued at around 50% (value of comfort). Even in this configuration the temperature has had a constant evolution, but with slightly higher values (around 18°C).

The Chart C refers to the entire period of monitoring. So from these charts shows that in order to obtain adequate conditions of comfort thermo-hygrometric in the environments hypogea of “Sassi di Matera” we must anticipate, already in the process of recovery, a adequate recycling and fan system of the environment more that an air conditioning system.

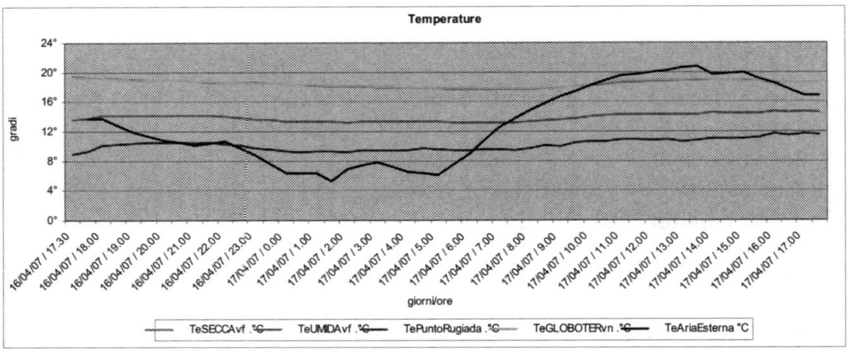

Chart C - Temperature of a rate of operation and use underground spaces.

Indeed assessing the UTA fairly compared to the size of the environment can to ensure acceptable levels of comfort, respectful of reference values established by legislation.

4 Analysis of the results

In conclusion, analyzing results obtained (Chart D) from measurements, we can make a critical and comparative summary between them, which may be briefly highlighted:

- Constant trend of internal temperature throughout the measurement period (13-15°C) for the hypogeum local not restored, 15°C the local hypogeum restored and 17°C for the local hypogeum restored and a rate of use; these values have been obtained against of the high thermal fluctuations daily with values exceeding 20°C (5°C night and daytime 28°C).
- The average temperature radiant of the walls has remained constant in the two configurations not restored and restored with value of (16°C), while the configuring restored and a rate of use had a slight increase reaching values of 18°C constant.
- The relative humidity has undergone large variation in the various configurations of local monitored. In fact, is has gone from high values,

approximately 95% for local hypogeum not restored, values of 65% for local restored and values of 50% for hypogeum restored and a rate of operation and use.

The last data even if it falls in the values established by rule, it is the upper limit of the range of tolerance for local with a similar use. This incident, however, is easily solved by setting a mechanical indoor air change of the domestic environments; in fact considering a replacement average volumes 5 volume/day the relative humidity inside drops to values around 50%, acceptable to the use of environment.

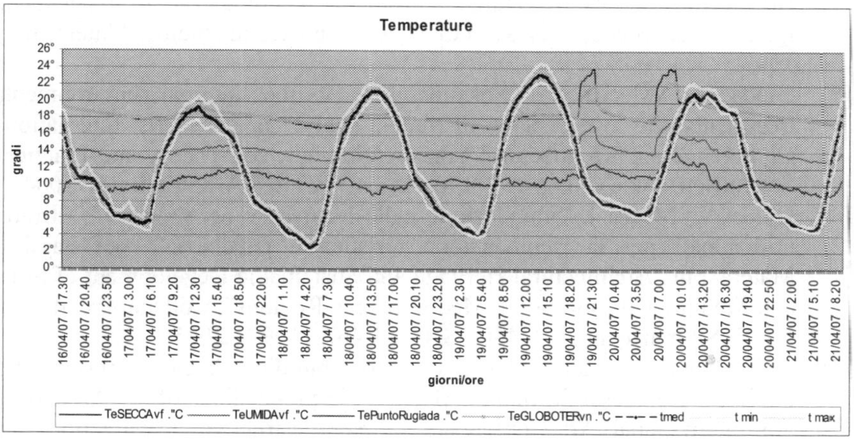

Chart D – Comparative chart.

5 Conclusions

The results of this evaluations shows that the parameters that govern the comfort "indoor" seem to assume, in the case of "Sassi di Matera", values respectful of recent regulations, although they should necessarily be repeated and extended in order to validate and to confirm the obtained data.

The research, therefore, shows how these architectures, realized with traditional techniques and materials, are an example of "sustainable construction", as can maximize the contributions in the solar cold season and minimise the hot season, taking advantage of the mass of the ground fly as thermal and encouraging natural ventilation of the environment. The "Sassi di Matera", therefore, seems to be almost a model of "bioarchitecture" in which natural stone, used "with wisdom", plays the key role of regulation. The study, in a further step, must systematize the data collected and deepen the study of "Sassi model" in order to reach the definition of methodological approaches and operational solutions to recovery the internal quality performance in similar contexts.

References

[1] Cotecchia V. Studio geologico-tecnico e stato di conservazione, in "Concorso Internazionale per la sistemazione dei Sassi di Matera", Matera, Italy, 1974.

[2] Cannarile. M., tesi di diploma, "Il risanamento da umidità dei Sassi di Matera: diagnosi e ipotesi di intervento su un ipogeo", Matera, Italy, 1995.

[3] Giuffrè A., Carocci C., Codice di pratica per la sicurezza e la conservazione dei Sassi di Matera, La Bautta, Matera, Italy, 1997.

[4] Buchicchio C., tesi di laurea, "Problemi tecnologici e di risanamento igienico nel recupero dei Sassi di Matera: un caso di studio", Matera, Italy, 1997.

[5] Guida A., Mecca I., "Sustainability of the internal environmental treatments: the case of Sassi of Matera (Italy)" in "The First International Conference on: Architectural Conservation between Theory and Practice", Scientific Book e CD pp 189–196, Dubai – United Arab Emirates, 2004.

[6] Guida A., Mecca I. "Innovazione nelle tecniche e nei processi costruttivi tradizionali per la riconversione turistica e residenziale dei Sassi di Matera", in "Architectural Heritage and Sustainable Development of Small and Medium Cities in South Mediterranean Regions", Collana Architettura, Ed. ETS Firenze, pp.443–458, ISBN 88-467-1199-8, 2005.

[7] A. Pagliuca, A. Guida, F. Fatiguso, "Stone building envelopes performance qualities: the "Sassi di Matera" (Italy)" - in proceedings of the ART2008 - 9th International Art Conference – "Non-destructive investigations and microanalysis for the diagnostics and conservation of cultural and environmental heritage", Jerusalem, Israel, 2008.

[8] N. Cardinale, A. Guida, F. Ruggiero, "Thermo-Hygrometric Evaluations in the Recovery of Rocky Buildings of the "Sassi of Matera" (Italy)" – Journal of THERMAL ENV. & BLDG. SCI., vol. 24-04-2001.

[9] N. Cardinale, F. Ruggiero, " A case study on the environmental measures techniques for the conservation in the vernacular settlements in Southern Italy" – BUILDIND AND ENVIRONMENT, 2002.

[10] G. Rospi, " Recupero architettonico e progettazione energetica delle architetture tradizionali mediterranee" – CODAT Meeting – Ancona, 2007.

Rock burst mechanics as a time dependent event

J. Vacek & S. Hrachová-Sedláčková
Klokner Institute,
Czech Technical University in Praha, Czech Republic

Abstract

This paper deals with the behaviour of open rock that occurs, for example, during long wall mining in coal mines, in deep tunnel, or shaft excavation. Long wall instability leads to extrusion of rock mass into an open space. This effect is mostly referred to as a bump, or a rock burst. For bumps to occur, the rock has to possess certain particular rock burst properties leading to accumulation of energy and the potential to release this energy. Such materials may be brittle, or the bumps may arise at the interfacial zones of two parts of the rock, which have principally different material properties. The solution is based on experimental and mathematical modelling. These two methods have to allow the problem to be studied on the basis of three presumptions:

- the solution must be time dependent
- the solution must allow the creation of crack in the rock mass
- the solution must allow an extrusion of rock into an open space (bump effect)

A part of the presentation will be bump video from mathematical and experimental tests.

Keywords: rock burst, bump, mining, rock mechanics, mathematical and physical modelling.

1 Introduction

The bump is the most dangerous event that can occur during excavation works. The surrounding rock is extruded into underground open space by severe force during a bump. This event may cause injury or even death to mining workers and it may be the cause of the destruction of the excavation space. This is the reason

WIT Transactions on the Built Environment, Vol 102, © 2008 WIT Press
www.witpress.com, ISSN 1743-3509 (on-line)
doi:10.2495/US080171

why the study of this problem is very important for theory and praxis (see [1–3, 7]).

Sufficient high pressure in the bump place is necessary for bump occurrence (usually great depth, but also tectonic pressure) and rock must be brittle and must have the disposition for a bump (properties that allow creation of bumps).

For the occurrence of bursts, mining velocity is also very important. In the same condition, when we excavate slowly, we give rock mass sufficient time to create cracks in the open space vicinity. This is the reason why stress concentration next to an excavation falls down, and a bump does not occur. If mining works proceed rapidly, the crack has no time to occur, and a bump appears. This feature is confirmed by old mining experience. This is also the reason why it is necessary to study this event as a time dependent problem.

A bump was studied for the case of a mine gallery inside a horizontal coal seam. Its mechanics and stress distribution on the top of seam was studied by mathematical and experimental modelling.

2 Experimental part

2.1 Testing devices

2.1.1 Loading cell

Fig. 1 shows the loading cell. It consists of a lower steel tank, which is designed for the horizontal forces caused by vertical load in araldite specimens. The loading cell is equipped with lucites on its sides, which allow observation of samples during the tests. The tank is shown in Figure 5. This loading cell models (simulates) the rock mass in the vicinity of the seam. We placed two araldite specimens in the loading cell (with dimensions of 160/400/70 mm), which model the coal seam. The gap between them corresponds to the width of a working gallery in a mine. We observed the mechanism and the history of coal bumps. The araldite specimen was covered with a soft duralumin sheet, and a force of meters were placed on it in the following manner: five comparatively thick force meters were placed near to its outer edge and another 15 thinner force meters were placed next to them, (see Figure 1 and Figure 5). In order to embed the

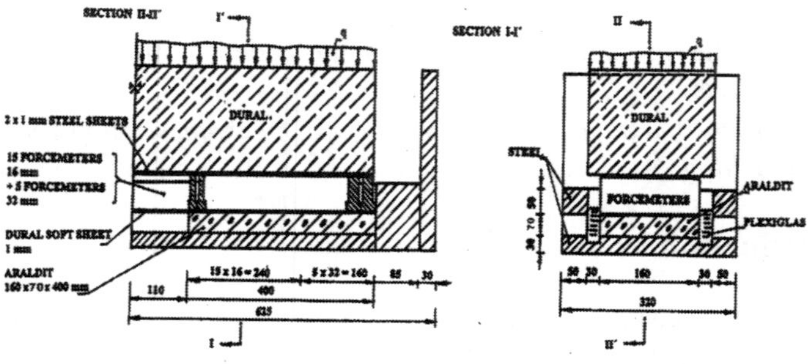

Figure 1: Scheme of loading cell.

force meters properly and to prevent them from tilting, another double steel sheet, 1 mm thick, was placed over the force meters. A block of duralumin with the height of 300 mm was placed over this sheet. This block simulates the handing wall and models stress distribution similar to that in reality (see Figure 2).

2.1.2 Force meters

Figure 5 shows the force meters. The force meters are 160 mm in length, 68 mm in height, and 16 or 32 mm in width. We can find four strain gauges on each force meter – two on one side 30 mm from the edge of the force meter and two on the other side 60 mm from the edge of the force meter. These allow us to measure the deformation along its full length.

Figure 2: Testing device. Two camcorders recorded the test.

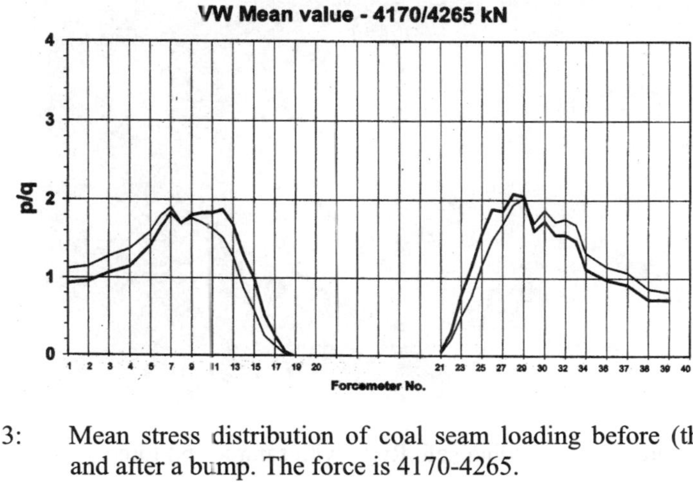

Figure 3: Mean stress distribution of coal seam loading before (thick line) and after a bump. The force is 4170-4265.

Dates of every force meter were read automatically by a Brűel and Kjaer strain gauge bridge every 10 seconds and deposited in a computer. A reading of 40 force meters takes 1,2 s.

2.2 Some of the results

More results are in [4–6].

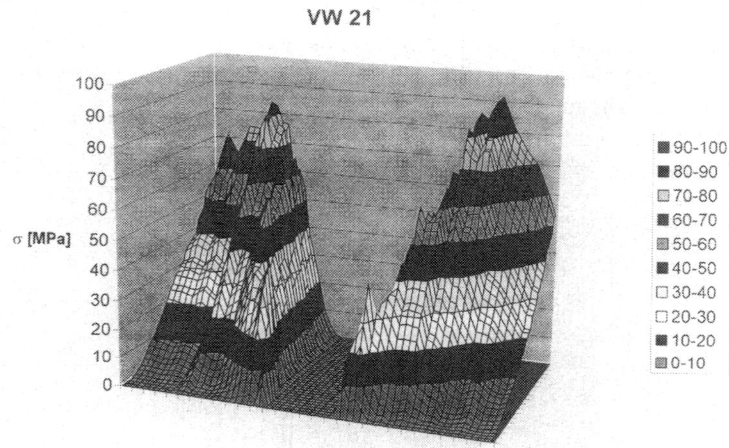

Figure 4: Loading of the coal seam during test VW 21 in axonometry. On axis x is the force meter number, on axis y is the time from the start of the test, and on the vertical axis is stress that acts in each force meter in corresponding time.

Figure 5: Loading cell. Right with sample and force meters.

Figure 6: State before bump.

Figure 7: First extrudes appears.

Figure 8: Early stage of bump.

Figure 9: Late stage of bump.

2.2.1 Mechanics of bumps

The last experiments were also recorded with a high-speed camera. It made it possible to watch bumps as a several second events. Figures 6–9 were chosen from continuous records. It follows from them that the studied bump started (similar to an earthquake) with a small extrude of rock mass, see Figure 6, followed by the main bump, Figures 7 and 8, and the state after the bump ended, Figure 9. The event sometimes ends with small extrudes of rock, but not in the described case. Video from the described case will be part of presentation.

2.2.2 Some facts, that follows from tests

- Extrude rock mass is located on the only narrow strip next to the free rock surface. This breadth was only 15–40% of the height of the coal seam and was proportionate to the bump force. The bigger the bump was, the broader the amount of extruded rock. The height of the coal seam was 70 mm in the experiment; the bump extruded from approximately 10 to 30 mm of rock. After the bump the surface of the coal was not smooth, it was uneven.
- In one case (one from about 30 cases) the bump came in two waves.
- Bumps in experimental conditions last about 20ms, (a two waves bump lasts 50ms) and the whole event lasts about 0,5s.
- The time between the first extrusion and the bump was 20–200 ms.

3 Mathematical model

PFC ^{2D} (Particle Flow Code in Two Dimensions) developed by Itasca, USA was used for the numerical modelling part of the project. A physical problem concerning the movement and interaction of circular particles may be modelled directly by *PFC*^{2D}. *PFC*^{2D} models the movement and interaction of circular particles by the distinct element method (DEM), as described by Cundall and Strack (1979).

By bonding two or more particles together, the particles of arbitrary shape can be created: These groups of particles act as autonomous objects, provided that their bond strength is high. As a limiting case, each particle may be bonded to its neighbour. The resulting assembly can be regarded as a "solid" that has elastic properties and is capable of "fracturing" when the bonds break in a progressive manner. *PFC* ^{2D} contains extensive logic to facilitate the modelling of solids as close-packed assemblies of bonded particles; the solid may be homogeneous, or it may be divided into a number of discrete regions of blocks.

The calculation method is a time stepping, explicit scheme. Modelling with *PFC*^{2D} involves the execution of many thousands of time steps. At each step, Newton's second law (force = mass x acceleration) is integrated twice for each particle to provide updated velocities and new positions, given a set of contact forces acting on the particle. Based on these new particle positions, contact forces are derived from the relative displacements for pairs of particles: a linear or non-linear force/displacement law at contacts may be used.

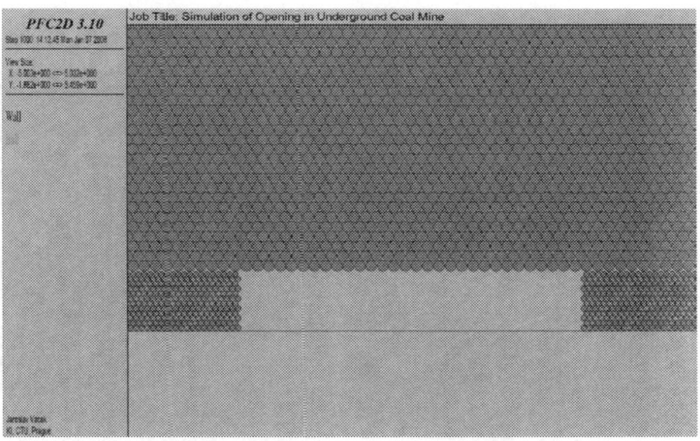

Figure 10: Mathematical model before a bump.

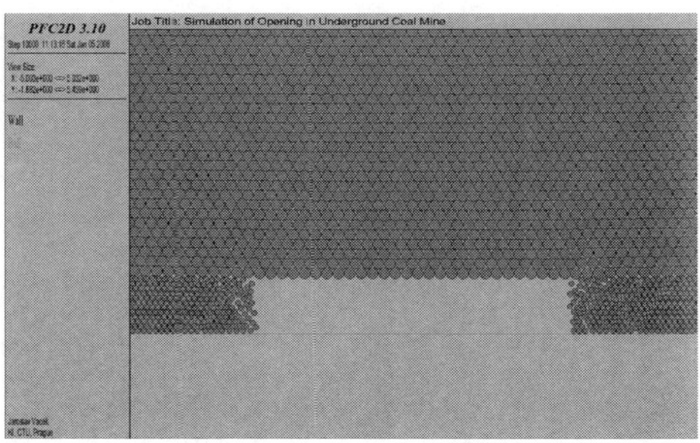

Figure 11: The first stage of a bump.

WIT Transactions on the Built Environment, Vol 102, © 2008 WIT Press
www.witpress.com, ISSN 1743-3509 (on-line)

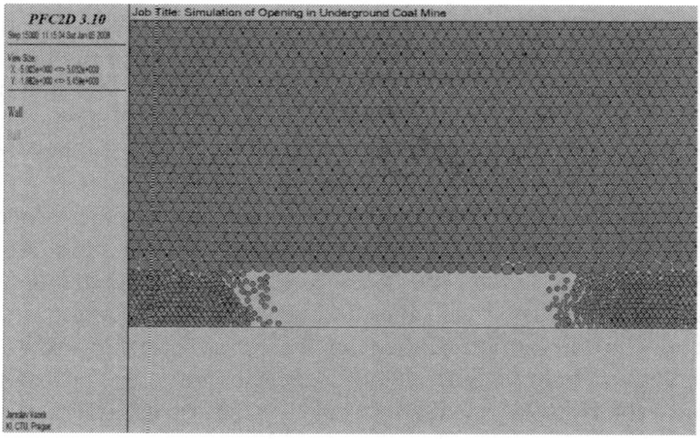

Figure 12: Next stages of a bump.

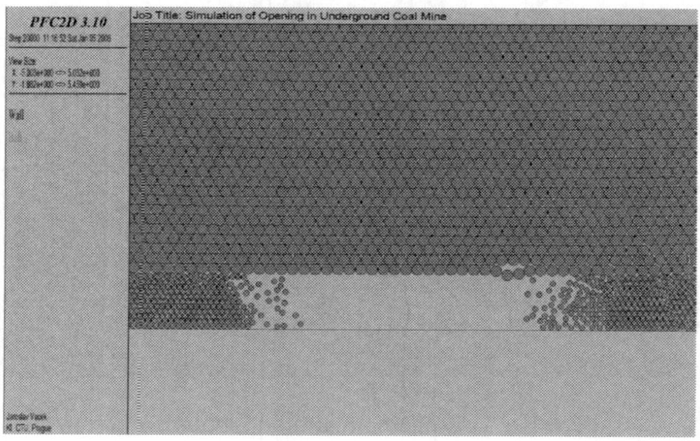

Figure 13: Next stages of a bump.

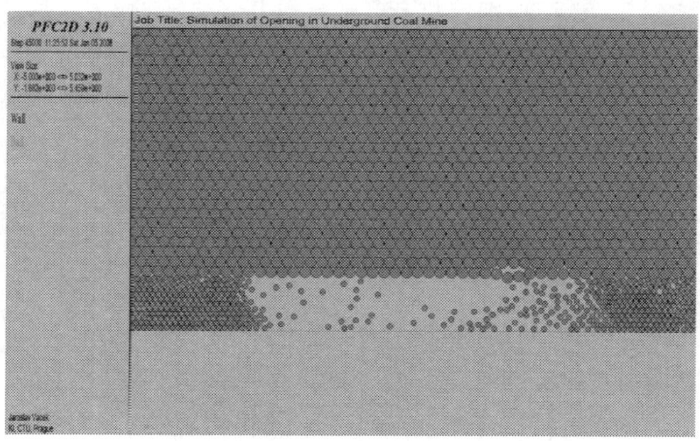

Figure 14: Late stage of a bump.

Figs. 10–14 show details of the mathematical model of the bump and Figure 15 shows the typical stress distribution along the coal mine. The stress grows (I, II) until the first bump initiation (III). This occurs at the 5th measurement cycle approximately, then the stress decreases in this location, but it increases simultaneously at the 7th measurement cycle, when the second bump initiation occurs (IV). The subsequent bump initiations can be expected in the 8th and 10th measurement cycles (V, VI).

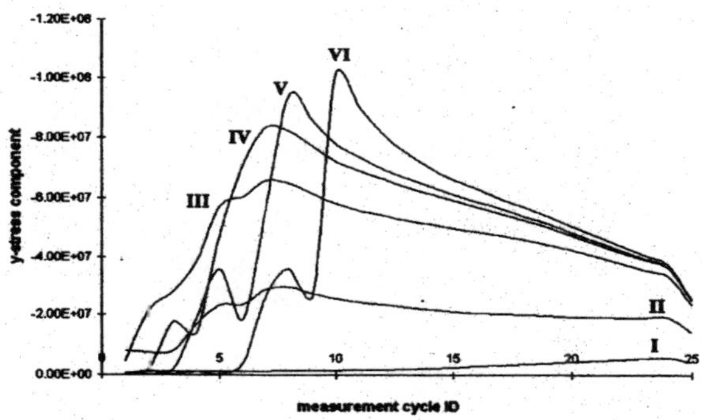

Figure 15: Stress distribution along the coal mine (left side) I - 5000 cycles, II-7000 cycles, III - 9000 cycles, IV - 12000 cycles, V - 15000 cycles, VI - 26000 cycles (number of time steps, iterations).

4 Conclusion

A combination of experimental and mathematical models appears very appropriate for a study of the stress distribution in a coal seam before and after bump initiation. Both methods enable a time dependent study of the problem and enable study of the development of cracks during bump initiation, and then extrusion of material into an open space during a bump. Thus they offer a description of the problem that is very close to reality.

Acknowledgement

This research and this paper have been sponsored by the Grant Agency of Czech Republic, (GAČR), grant number 103 / 08 / 0922 “Influence of shocks and impacts loading on structures”.

References

[1] Foss, M., M., Westman, E., C.: *Seismic Method for in-seam coal mine ground control problems,* SEG International Exposition and 64th Annual Meeting, Los Angeles, 1994, p 547–549

[2] Goodman, R. E.: *Introduction to Rock Mechanics,* John Wiley & sons, 1989, 562 p.

[3] Toraño, J., Rodríguez, R., Cuesta, A.: *Using experimental measurements in elaboration and calibration of numerical models in geomechanics*, Computation Methods and Experimental Measurements X, Alicante, 2001, p. 457–476

[4] Vacek, J., Procházka, P.: *Rock bumps occurrence during mining*, Computation Methods and Experimental Measurements X, Alicante, 2001, p. 437–446

[5] Vacek J., Bouška, P.: *Stress distribution in coal seam before and after bump initiation*, Geotechnika 2000, Glivice- Ustroň 2000, p. 55–66

[6] Vacek, J., Procházka, P.: *Behaviour of brittle rock in extreme depth*, 25th Conference on Our World in Concrete & Structures, Singapore, 2000, p. 653–660

[7] Williams, E. M., Westman, E. C.: *Stability and Stress Evaluation in Mines Using In-Seam Seismic Methods,* 13th Conference on ground control in mining, US Bureau of Mines, 1994, p. 290–297

Spatial organization and economic analysis in sustainable transit oriented development

N. Mohajeri
Department of Art and Architecture, Islamic Azad University, South Tehran Branch, Iran

Abstract

Transit oriented development (T.O.D.) is one of the most important subjects in sustainable urban development. This strategy has been taken for decreasing the journey demand especially combining land use and transportation development with focus on the land development around railway stations. According to this statement, the provision of a station complex is considered as an option for approaching a sustainable urban transportation.

This paper analyses the spatial organization underground stations after a brief defining of transit oriented development. This paper also is going to discuss the benefits of a station complex which is made according to sustainable development with focus on the economical effects.

Keyword: spatial organization, sustainability, transit oriented development.

1 Introduction

In order to get into economic development, the existence of an inner city rail transportation network is considered an indispensable tool for success. Since three decades ago forward looking urban designers and urban planners in connection with intelligent development in cities, in order to minimize demands for daily inner city trips, have put great emphasis on the coordination of land use and transportation systems. This idea nowadays clearly implies concentration of various daily activities in and around underground or city railway stations. This impact of such magnitude particularly in areas immediately next to the stations is quite evident and by passage of time every station creates a magnet of its own consequentially ending in gathering up of an assortment of utilities and intense

WIT Transactions on the Built Environment, Vol 102, © 2008 WIT Press
www.witpress.com, ISSN 1743-3509 (on-line)
doi:10.2495/US080181

built up around it all relative to the station's position in the city and on the zonal level.

Being next to and near the stations facilitates easier access for residential, commercial, and service centers while simultaneously helping to save on time and expenses. The effects of such advantages in other areas are best reflected in increasing of the value of adjoining properties and that means there is possibility of gaining new revenues which can be spent on construction and maintenance of rail transport systems. Our article is concentrating on an appraisal of underground economic impacts on the development in areas situated around stations' circle of influence.

2 History and definition

During the Second World War in the United States low density built up and development plans inclusive of non centralized utilities formed the method of land use in cities. Such a manner of urban development and planning encouraged an amazing growth in the number of personal and commercial journeys during those years. In recent years the impacts of rising volume of journeys and rapid and extensive expansion of cities have created the problem of air and sound pollution as well as the rise in the level of greenhouse gases. Now urban planning organizations and transport agencies are taking up policies and strategies with regard to new approaches to the patterns of built up density and appropriate usage of land.

Calthorpe in 1993 presented an elaborate picture of transport oriented urban development. He has defined transport oriented development or TOD as a space or particular limit consisted of high density residential built up, small shops, administrative and commercial complexes and other utilities. Stores and service centers are situated in its commercial core where on-foot movement up to 600 meters (or 10 minutes walk) to places of residence is feasible. The transport station is right in the center. Utilities inside the core are totally horizontal. Apartments and offices are situated on upper floors above shopping centers.

Other areas with lower density are within 600 and 1600 meters distance from the core .These areas could be used for locating residential suites of various sizes, small parks, schools and some light industries. In the design of houses the emphasis is on a mixture of traditional and modern characteristics in such a way that their easy access to out street parking could be available. Streets in the vicinities are mainly corresponding to the network pattern and facilitate access to the core by foot or bicycle.

3 Principles observed in implementation of station complexes

3.1 Space planning in transit stations

The principal users of station complexes are pedestrians who travel on foot. All inner city journeys also includes elements of walking whether in the beginning or at the end of a journey. Therefore space planning for a station complex around light railway lines has to be in such a way that on foot pedestrians could make an effective use of it, in other words the maximum distance from the station should not exceed a walk of 5 to 10 minutes or a distance of 600 meters.

3.2 Land use design according to transit oriented

The design of land use in areas adjacent to light railway transport stations has to have the following characteristics:

- High built up density in working and residential environments
- Moving journey time to other non peak hours
- Encouraging the public to use light city train and roads in their journey routes
- Encouraging the distribution of non working hours during the day and the week
- Creating motivation among on foot pedestrians to use stations

The various utilities located in this limit includes the following items:

- High schools and higher education institutions
- Residential areas with average and high built up density
- Small shops
- Service centers
- Cultural and entertaining centers

3.3 Land use combination

The space allocated to a station has to provide favorable conditions for combining residential centers, offices, shopping and service centers. The combination of land use could be of horizontal or vertical arrangement in such a way that several utilities are located in multi storied buildings or can be scattered in various adjacent buildings. The importance of this is derived from the fact that in the influence circle of a station it is possible to cover various needs through on foot movement.

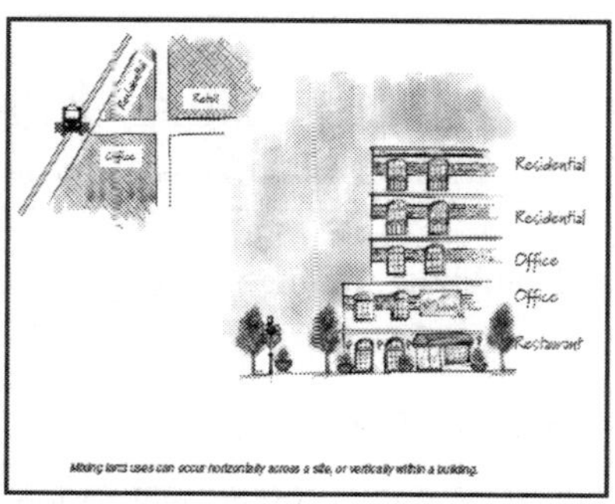

Figure 1: Combined land uses around a station in horizontal, vertical arrangements and within buildings, floors.

3.4 Restraining land uses when not harmonious with transportation system

Since the major emphasis of station complexes is on transfer transportation and pedestrians, it is necessary to prevent the spread of land uses which encourage the use of personal vehicles within station's sphere of influence. Such utilities are as follows:

- Utilities occupying large land with low build up density
- Utilities requiring large space for guaranteeing the prevention of damages to the surrounding environment
- Utilities requiring the use of private vehicle
- Utilities which impose limitation on foot access

Some examples of these utilities are as follows:

Vehicles, centers of service and repair/ centers of sale and purchase of vehicles/carwash centers/fuel stations/public parking/small industries/goods warehouses/low density commercial centers/low density residential areas.

Such areas should not be within immediate reach of station space and the best option is keep them off the circle of station's planning domain as much as possible.

3.5 Improving built up density around station

Built up density around transit stations should be increased or in other words such density must be affected in places where the best access to the rapid transportation system is available. Furthermore, high density utilities and buildings should rather be located in the nearest points to the city railway station building.

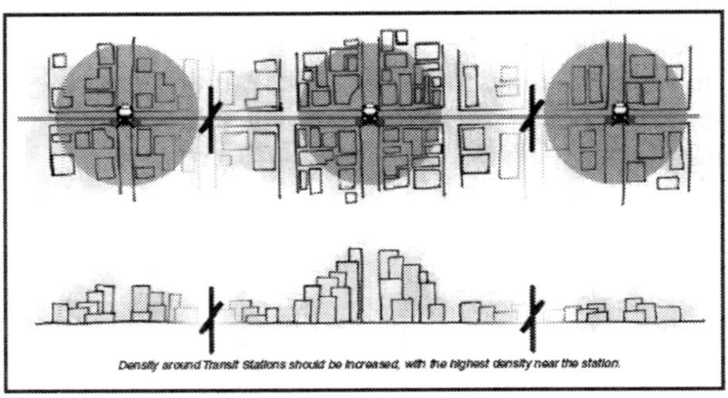

Figure 2: Utilities and buildings of high density near city railway station building.

3.6 Minimizing the impacts of density

With the increase in density around stations inclination towards living near centers situated off this limit grows and therefore some solutions to confront such phenomenon have to be found.

3.7 Designing good quality routes for pedestrians

A trouble free and suitable route for making connection by on-foot movement should possess the following characteristics:

Shortness/ continuity/ obstruction free/ safety/ plainness and clear definition/ design based on climatic condition of the city (for example the necessity of sheltering people from sharp sunshine).

Figure 3: Harmonizing general built up density in surrounding area with station complex.

3.8 Compact development structure

The placement of buildings in a compact space should be managed in such a way that any access to them for pedestrians be easy either by main passages or less prominent passages needless of long walk from one building to another.

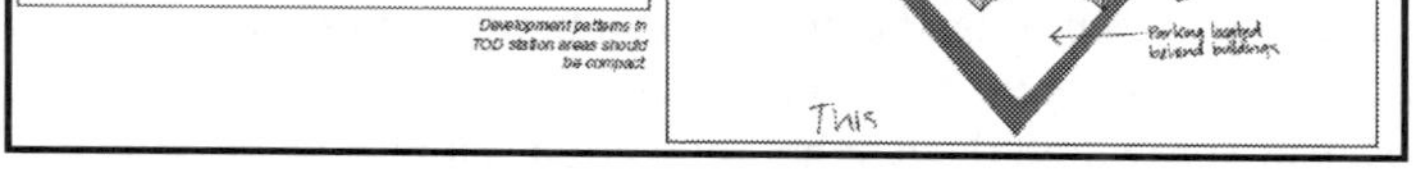

Figure 4: Compact development pattern for developing station's surroundings.

3.9 Taking into consideration the climatic characteristics of all seasons

In designing main walking routes and also transport facilities, sheltering devices to protect people from climatic phenomenon's (sun rays, rain, snow etc…) should be taken into consideration. This purpose could be reached by the way of constructing closed spaces in stations and covered porches outside.

3.10 Placing parking lot in right spots

The spaces given to parking lots should not hinder the comfortable and safe movements of pedestrians within station. The major parking must be accessible by main roads and other passages in such a way that they do not come into contact with pedestrians, routes. The pedestrian's passages which may lead to transit station, main administrative spaces, and high density residential areas must be separated from parking spaces. Parallel to pedestrian's passages leading to transit station, parking lots should be designed and constructed in such a manner that passing vehicles traffic could be kept at its minimum in the routes of walking public.

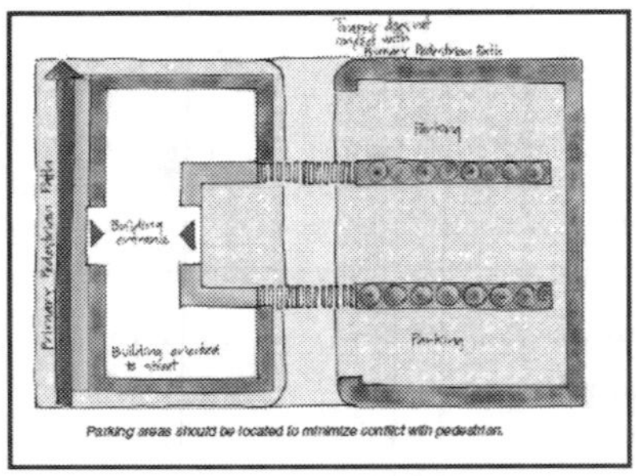

Figure 5: Appropriate placement of parking lots around station with regard to the pedestrian's passageways.

4 The goals of constructing station complexes

The project is concerned with planning for areas adjacent to station based on policies and methods of materializing the following key goals:

- Optimum use of lands around station

An optimum use of station surrounding land will support the interests of passengers as well as providing groups of local people with a place for launching various activities. The other positive effect of station complexes is to make ground for redistribution of land exploitation among sections devoted to services, employment and building activities in the city.

- Density growth in the neighborhood of a station

Growth in the level of density around stations with regard to resultant passenger increase would entail rise in efficiency of stations and would support rapid transportation services.

- A pedestrians oriented design

The station complexes would maintain an appropriate, comfortable, straightforward and safe connection among pedestrians in pursuit of supporting pedestrian friendly stations and to create motivation for choosing public transportation as the means of commute.

- Changing every station and its vicinity into an attractive place for pedestrians

Due to existence of such facilities in the limits of station complexes, every space in a station could become an appealing and unique environment in the eyes of on-foot visitors and consequently offer revenue generating potentials of considerable importance.

- Parking management and the management of vehicle transportation

Designs and architectures in the limits of station complexes are planned in such a way that would not allow transportation systems like bus system, private vehicles and parking to hinder the movement of on-foot pedestrians and damage their heath by creating pollution.

- Planning in local communities

Transportation oriented urban development has to produce some advantages for the benefit of local residents. These advantages could consist land use increase, appropriate services, building activities of various types, increase in the number of transport service options and more compatibility of walking environment.

5 The benefits of constructing station complexes

5.1 The benefits of constructing station complexes based on principles of continuous development

The benefits of station complexes based on principles of sustainable development are shown in the following model:

Table 1: The benefits of station complexes based on principles of sustainable development.

Society	**Economy**	**Environment**
Improving journey options via widening the variation of vehicles Increase in employment, service offering and…. Balancing periods of presence at home and work place. The health advantages of walking in the designed areas.	Maximum exploitation of rapid transportation infrastructure. Decrease in the level of traffic and the consequent expenses. New formation of obsolete or misused industrial and commercial environment.	Decrease in the level of greenhouse gases Decrease in consumption of energy by optimum exploitation of land and transportation Upgrading the quality of air.

5.2 Environmental impacts

If we study the question of traveling by private vehicles from the point of view of energy consumption we realize that the volume of VMT distinctive in a direct way affects the environment of concerned area. The residents of area recognize and understand fully the fuel consumption related symptoms such as air pollutions; air pollution oriented illnesses as well as negative effect of air pollution on the quality of water in the area.

Studies show that TOD negative effects in all times are close to 46% less than those related to traditional systems of transportation. These effects such as fuel consumption, carbon dioxide and carbon black articles environmentally affect the health of citizens and damage the beauty of the city.

Dissemination of carbon dioxide as a result of fuel burning create urban pollutions beyond urban areas leaving negative impression on pastures and the climatic conditions. The negative effects left by carbon dioxide on pastures equals in value the stretch of woodland which has to be destroyed in order to produce fuel. Illustrating negative effects of carbon dioxide in this manner would show more clearly negative effects generated by its dissemination through burning of fossilized fuels.

5.3 Economic impacts

Station complexes produce many economic impacts in societies and cities at large. They create extensive variations in residential units which correspond to the needs of different generations and their incomes. It also could act as catalyst for ever more development of city and the economy. Executors of TOD projects benefit from reduction of occupied lands which would otherwise be used as car parking lot. On the other hand the residents of the area also benefit from reduction of transport costs and consequent rise in their power of spending.

5.3.1 Short term impacts

- Saving: Such benefit mainly comes from providing services to greater number of people in a smaller area.
- Savings generated by suitable location: Such benefits are gained by creating opportunity for greater number of people to have access to essential facilities.

In addition to financial gains for local government, also by expansion of such a well planned transportation system the following benefits are accessible:

- Rise in prices of residential lands due to new urban management
- Rise in prices of residential lands due to accessibility of comfort, advanced public transit system and cheaper services.

5.3.2 Long term impacts

- Growth in the value of regional assets due to management's appropriate policies which create ground for rise in the revenues derived from tax and land value increase
- Rise in average workforce efficiency

- Rise in general efficiency facilitated by creation of high density cities and use of efficient systems of transportation.
- Creating a greater share of regional revenues relative to those of other regions by optimum resource management
- No guarantee could be given for full realization of all above mentioned economic benefits but there is no doubt even the realization of even a small portion of these benefits could have considerable impacts on any region.

5.4 Appraisal of benefits and impacts of utilizing station complexes

An appraisal of benefits and impacts of TOD could be studied in the form of two main groups. The following table shows the positive impacts of TOD for governments and societies:

Table 2: The positive impacts of TOD for governments' societies.

Societies	Government
Increase in accessibility of various areas by creating byways	Rise in the number of public transportation users
Improvement in condition of local traffic	Rise in revenues of transport orientation
Rise in high density infrastructures and variation in usage of lands which help to make optimum use of lands	Rise in revenues gained from tax and profit of urban land development
Development in economy by methods such as increasing the level of employment	Reinforcement of organizational relations

6 Conclusion

As explained in the paper the limits of surrounding area around inner city stations right from the first phase of implementation could be defined as the area of change which in various times relative to created capacities would take in a new degree of density and land usage. It is necessary by taking into account the goal of logical utilization of created opportunity such changes be settled and managed realistically. The recognition of opportunities and capacities for organization and planning what inevitably would take place in future is necessary and unavoidable.

Investment in city railway projects would create ground for various changes such as economic developments within stations, circle of influence. These developments in a chain of time would bring about rise in the values and also would help many changes to appear on the face of urban life. In addition to that, investment in the city railway projects would set off space rearrangement of areas adjacent to the stations.

References

[1] J.M. Pertchik & A.R. West., *TOD initiative west Palma Marlette and feasibility study Report*, Treasure coast Regional Planning Council, October 2004.

[2] S. Polikov., *Leander TOD market Analysis*, Capital Market Research, INC, Jan. 2005.

[3] Costs of Sprawl *Transit Cooperative Research Program*, 2002.

[4] Norman Y. Mineta *A New Planning Template For TOD, 2004*

[5] Robert Cervero *Accessible Cities and Regions A Framework for Sustainable Transport and Urbanism in the 21st Century*

[6] Peter Newman *Transit Oriented Development an Australian Overview, 2005*

[7] Kathi Holt-Damant *Transit Oriented Development*, Making Happen, 2002.

[8] Peter J. Marcotullio *The compact city environmental transition theory and Asia-Pacific urban sustainable development, 2004*

[9] Walter Siembab *Technology as Part of Transit Oriented Developments, 2004*

[10] Richard Willson *Parking Policy for TOD* California State Polytechnic University Pomona, 2002.

[11] Michael Gilat *Coordinated Transportation and Land Use Planning in the Developing World – The Case of Mexico City, 2004.*

[12] Ivonne Audirac *Marketing Transit Oriented Design* The Florida Department of Transportation Public Transit Office 2004.

[13] Shaoming Zhang *Feasibility Study on Transit-Oriented Development* Using Urban-Form and Non-Urban-Form Variables, 2005

The effect of a baffle on the heat transfer in underground auxiliary ventilation systems

S. M. Aminossadati[1] & B. Ghasemi[2]
[1]*CRCMining, School of Engineering, The University of Queensland, Australia*
[2]*Engineering Faculty, Shahrekord University, Iran*

Abstract

This paper presents a computational fluid dynamics simulation to examine the effects of the baffle on the air flow field and the heat transfer rate in a space located beneath a ventilation channel. The computational model consists of a two-dimensional channel equipped with a thin baffle hanging from the top wall and a cavity located underneath simulating the underground space. Air flows trough the channel at a uniform velocity, u_c, and temperature, T_c. The cavity base is assumed to be heated at a constant temperature, T_h, while the vertical walls of the cavity are well-insulated. The top wall of the channel is maintained at a constant temperature, T_c. The continuity, momentum and energy equations are solved numerically using the control-volume approach to examine a combination of natural and forced convection flows. The results show that even though the existence of the baffle improves the cavity ventilation performance at low Richardson numbers, it has a negative impact at high Richardson numbers.
Keywords: heat transfer, forced convection, underground ventilation, baffle.

1 Introduction

Nowadays, underground spaces are being utilised for a wide variety of purposes such as sport halls, power stations, waste repositories, mining operations and underground cities. The ventilation system is a key factor in the design of such spaces. Moreover, as the underground operations are continuing to develop to greater depths, the associated issues with the ventilation systems such as the heat load and airborne contaminants are becoming more important.

WIT Transactions on the Built Environment, Vol 102, © 2008 WIT Press
www.witpress.com, ISSN 1743-3509 (on-line)
doi:10.2495/US080191

Convection heat transfer in channels with open cavities has been commonly considered in the thermal design of underground spaces. A review of the literature indicates numerous studies on the flow field and heat transfer analysis of channels with open cavities Channels with open top upright cavities have been used by some researchers to simulate the heat transfer mechanism for either pure natural convection [1–5] or mixed convection [6–8]. Recently, Manca et al. [9] presented a numerical study on mixed convection in an open cavity with a heated wall bounded by horizontally insulated plate. They considered three basic heating modes for the open cavity: Assisting flow mode, opposing flow mode and heating from below mode. It was demonstrated that opposing flow mode had the highest thermal performance in terms of both maximum temperature and average Nusselt number. Brown and Lai [10] considered a horizontal channel with an open cavity and studied the combined heat and mass transfer numerically. In their study, only the bottom wall of the cavity was heated at a constant temperature and the other walls of the cavity were insulated. Correlations were obtained for combined heat and mass transfer which covered the entire convection regime from natural, mixed, to forced convection. Leong et al. [11] studied the mixed convection for the same geometry used by [10], showing that the Reynolds number and Grashof number control the flow pattern and the occurrence of recirculating cells while the cavity aspect ratio has a significant influence on the orientation of these cells. The transition to the mixed convection regime was a function of the relative magnitude of Grashof and Reynolds numbers. They argued that in the mixed convection regime, the heat transfer was reduced and the flow might have become unstable. Manca et al. [12] presented an experimental investigation on the mixed convection in a channel with open cavity having a heated wall on the inflow side. The geometry of the problem was similar to the one used by [9]. The results showed that the maximum dimensional temperature rise values decreased as the Reynolds and Richardson numbers decreased. For Re=1000, two nearly distinct fluid motions were observed: a parallel forced flow in the channel and a recirculating flow inside the cavity. For Re=100, the effect of a stronger buoyancy determined a penetration of thermal plume from the heated plate wall into the upper channel.

Auxiliary ventilation is often used in underground spaces to manage dust, gas and heat. Critical factors to auxiliary ventilation include the condition of supply air and the generation of dust, gas and heat in the underground space. The capacity of auxiliary ventilation systems is typically limited by the characteristics of the ducting system. A conductor baffle is often used in the ventilation channels to direct a portion of the ventilation air into a designated underground space. As such, the objective of this research is to study the effects of a baffle on mixed convection heat transfer in a two-dimensional horizontal channel with an open cavity.

2 Problem definition

The geometry considered here is a two-dimensional horizontal channel with an open cavity located at the bottom of the channel. Fig. 1 shows a schematic diagram of the channel. The channel is equipped with a thin baffle hanging from the top wall. Air is introduced into the channel at a uniform velocity, u_c, and temperature, T_c. The base of the cavity is at a high temperature, T_h, while the vertical walls of the cavity are insulated. The top wall of the channel is maintained at a low temperature, T_c, and the bottom wall is thermally insulated. The height of the inflow and outflow openings is equal to the depth of the cavity. This model can be used in the study of the effects of the baffle on the channel thermal performance, at various Richardson numbers.

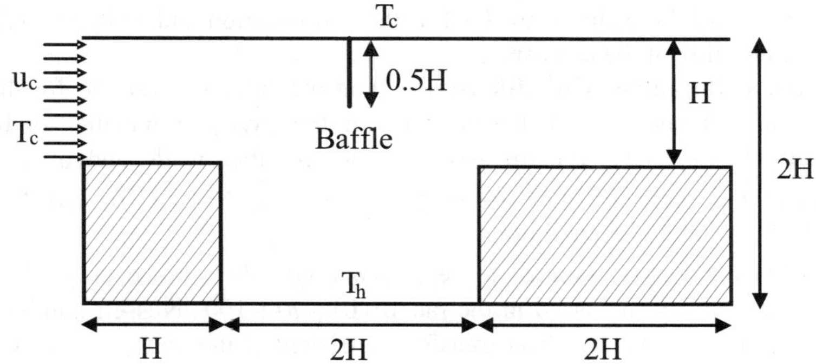

Figure 1: A schematic diagram of the model geometry.

3 Governing equations

By assuming, laminar, incompressible and two-dimensional convection fluid flow with constant properties except for the fluid density in buoyancy forces, the non-dimensional continuity, momentum and energy equations can be written as follows considering Boussinesq approximation:

$$\frac{\partial U}{\partial X}+\frac{\partial V}{\partial Y}=0 \tag{1}$$

$$\frac{\partial U}{\partial \tau}+U\frac{\partial U}{\partial X}+V\frac{\partial U}{\partial Y}=-\frac{\partial P}{\partial X}+\frac{1}{\mathrm{Re}}\left(\frac{\partial^2 U}{\partial X^2}+\frac{\partial^2 U}{\partial Y^2}\right) \tag{2}$$

$$\frac{\partial V}{\partial \tau}+U\frac{\partial V}{\partial X}+V\frac{\partial V}{\partial Y}=-\frac{\partial P}{\partial Y}+\frac{1}{\mathrm{Re}}\left(\frac{\partial^2 V}{\partial X^2}+\frac{\partial^2 V}{\partial Y^2}\right)+\frac{Gr}{\mathrm{Re}^2}\theta \tag{3}$$

$$\frac{\partial \theta}{\partial \tau}+U\frac{\partial \theta}{\partial X}+V\frac{\partial \theta}{\partial Y}=\frac{1}{\mathrm{PrRe}}\left(\frac{\partial^2 \theta}{\partial X^2}+\frac{\partial^2 \theta}{\partial Y^2}\right) \tag{4}$$

In these equations, the steady-state solutions can be obtained by setting the time dependence terms to zero. Based on the Boussinesq approximation, density is assumed to be constant in all the terms except for the buoyancy term. In the above non-dimensional equations, τ is the time in a dimensionless form ($\tau = u_c t/H$), all the lengths are normalised by the depth of the cavity ($X = x/H, Y = y/H$). Velocities are normalised by the inlet velocity ($U = u/u_c, V = v/u_c$). The pressure is normalised as $P = \bar{p}/\rho u_c^2$, where $\bar{p}$ is the modified pressure ($\bar{p} = p + \rho_C g y$). The non-dimensional form of temperature is $\theta = (T - T_c)/(T_h - T_c)$. Fully-developed conditions are considered at the exit section of the channel.

Hydrodynamic Boundary Conditions: No slip boundary conditions (U=V=0) on all the walls, V=0, U=1 at the entry section and V=0, $\partial U/\partial X = 0$ at the exit section of the enclosure.

Thermal Boundary Conditions: $\theta = 0$ at the entry section and for the top wall of the channel. $\theta = 0$ for the base of the cavity. Adiabatic conditions ($\partial\theta/\partial X = 0$ or $\partial\theta/\partial Y = 0$) are assumed for all other walls and at the exit section of the enclosure. For the unsteady analysis, U=0, V=0, and θ=0 are considered as the initial conditions.

Grashof number is assumed to be constant $Gr = 10^6$; however, Richardson number, Ri=Gr/Re^2, is varied in the range $0.01 \le Ri \le 100$. Nusselt number, Nu, is a measure of convective heat transfer coefficient at the bottom of the cavity. The local Nusselt number is defined as $Nu = -(\partial\theta/\partial Y)_{Y=0}$ The average Nusselt number, Nu_m, is obtained by integrating the Nusselt number over the bottom wall of the cavity.

4 Numerical method

The system of governing equations (1-4) with the above-mentioned boundary conditions is solved through a control volume formulation of the finite difference method. The well known SIMPLE algorithm is used to handle the pressure-velocity coupling. The convective fluxes across the surfaces of the control volume are determined by the power law discretisation scheme. A program code in Fortran is developed to follow the algorithm. In order to select the appropriate grid, a systematic grid independence study is carried out and the effect of grid refinement on the flow parameters is examined. According to the results of the grid independency study and for the sake of computation time, a equidistant grid of 110×40 is selected for the analysis.

5 Results

As the first step, the validation of the computer code for the mixed convection in the open cavity was assessed. This was done by running the present code with the conditions used by Manca et al. [9]. They considered an open cavity in a channel with a uniform heat flux on the vertical wall. Fig. 2 shows the variation of the minimum stream function with Richardson number. As can be seen from this figure, there is a good agreement between the two studies.

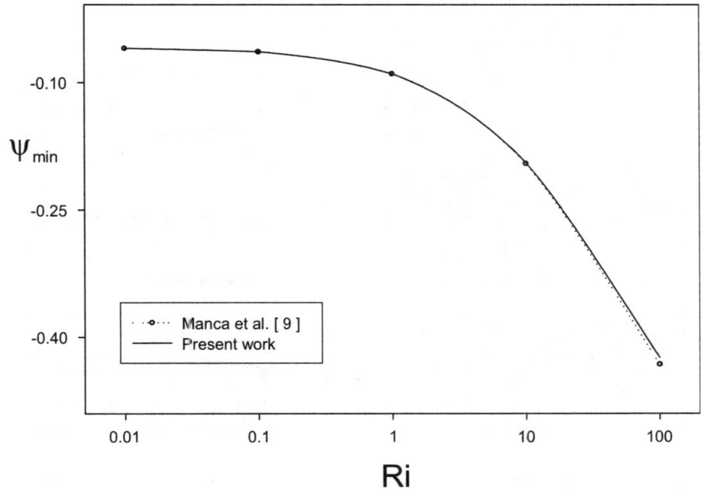

Figure 2: A comparison between the present work and Manca et al. [9].

The second step was to study the effects of baffle on the flow and temperature fields and the heat transfer rate. To accomplish this study, the following values are used in the analysis. $\Pr = 0.71$, $Gr = 10^6$, and $0.01 \le Ri \le 100$. The flow and isotherms are shown in Fig. 3. Streamlines and isotherms are presented on the left and right hand sides of the figure, respectively.

Plots presented in Fig. 3(a) correspond to three different values of the mixed convection parameter Ri=0.1, 1, and 10 at a channel with no baffle, whereas the plots in Fig. 3(b) are for a channel in which a baffle is installed. Fig. 3(a), shows that at Ri=0.1, the airflow does not have any tendency to enter the cavity and as a result, an insignificant circulation flow is evident in the cavity. As the Richardson number increases, stronger buoyancy cells appear in the cavity and begin to cover the whole cavity. It can be said that, an increase in Richardson number ($Re=Gr/Re^2$) is associated with a decrease in Reynolds number since Grashof number is assumed constant. Consequently, the externally induced airflow in the channel become less controlling and that associates with the onset of natural flow circulations. In the next scenario, as seen in Fig. 3(b), the baffle directs some portion of the external flow into the cavity and therefore circulation cells are generated in the entire cavity at low Richardson number of Ri=0.1,. The isotherms also show an improvement in the penetration of the external flow into the cavity when the baffle is used.

Fig. 4 shows the vertical velocity component of the airflow at the horizontal midsection of the cavity, Y=0.5,. The plots for both channels with and without baffle are presented in this figure. For small Richardson numbers, the peak of the velocities in the channel without baffle is almost flat showing only a slight buoyancy effect. In the cavity of the channel with baffle, the recirculation zones appear at low Richardson number (see Fig. 3(b)) and sharp changes in the slope of the velocity profiles observed near the vertical walls of cavity. As the

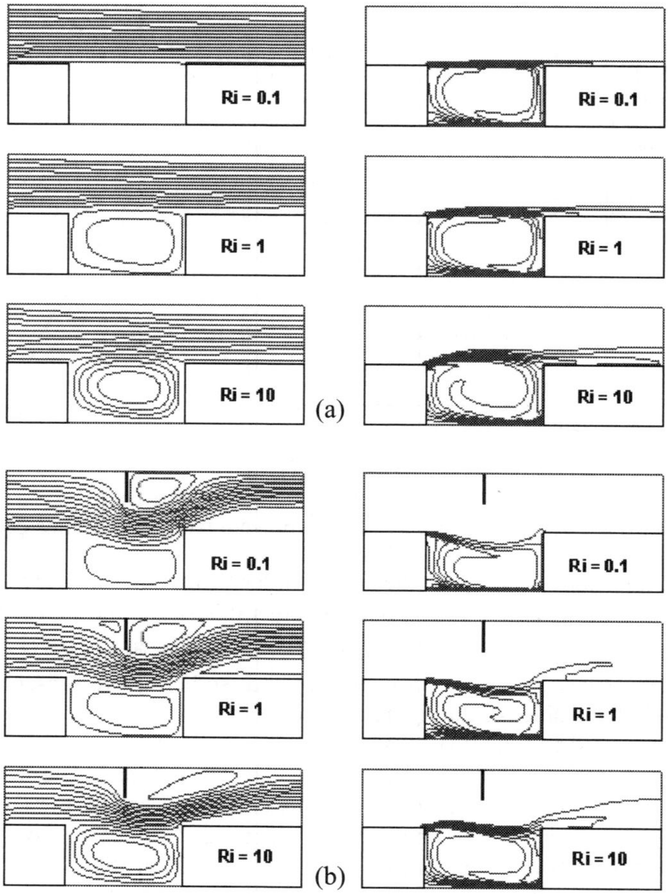

Figure 3: Streamlines (left) and isotherms (right) at various Richardson numbers (a) channel without baffle, (b) channel with baffle.

Richardson number increases, the recirculation zones become stronger in both cavities and the peak of the velocity profiles moves upwards. In this case, only a slight difference is evident between two channel profiles.

The variation of average temperature of the entire cavity is presented in Fig. 5. The results show that for the channel with no baffle the average temperature does not change significantly with low Richardson numbers. In addition, it can be seen that at Richardson number of Ri=0.01, the average temperature for the channel with baffle is much lower than that for the channel without baffle and that is because of the baffle directs some portion of the dominated external air flow with a relatively lower temperature into the cavity. As Richardson number increases, the average temperature increases because the buoyant flow becomes stronger. An interesting point is that at Ri=100, the average temperature for the channel with baffle is higher than that for the channel without baffle. This will be explained in the subsequent sections.

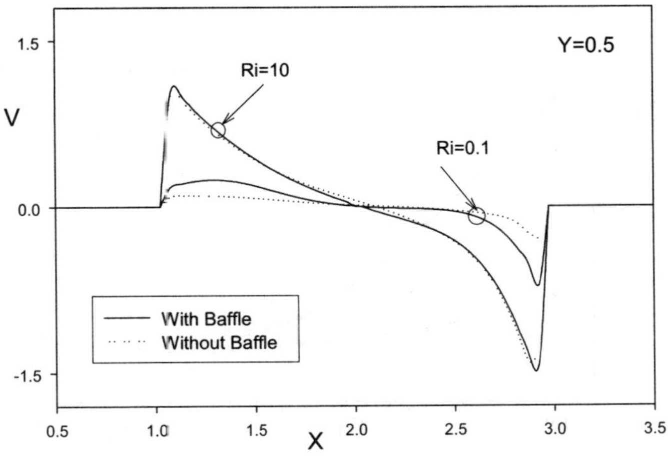

Figure 4: Vertical component of velocity at midsection of the cavity.

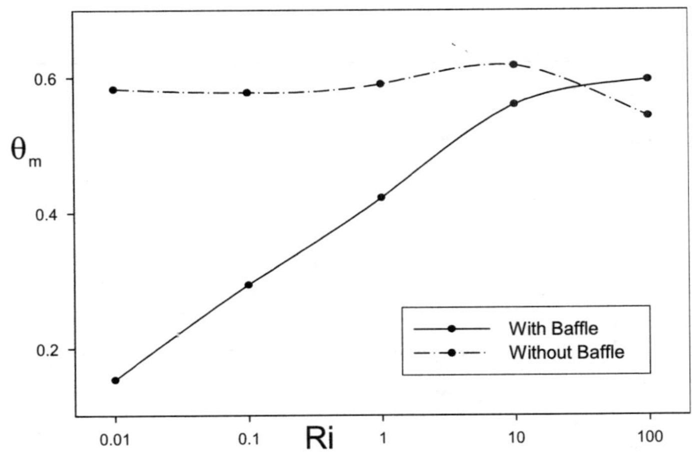

Figure 5: Average air temperature in the cavity.

Heat transfer rate from the cavity in terms of the variation of average Nusselt number for the two channels with and without a baffle is presented in Fig. 6. As expected, at low and medium Richardson numbers, the cavity of the channel with baffle has a better thermal performance.

A better comparison of the heat transfer performance of the two different channels is presented in Fig. 7. Here, the variations of the average fluid temperature ratio, θ_{mB}/θ_m, and the average Nusselt number ratio, Nu_{mB}/Nu_m are shown. The results of the average Nusselt number ratio shows that at Ri=0.01, the heat transfer rate for the channel with baffle is approximately three times higher than that for the other channel.

WIT Transactions on the Built Environment, Vol 102, © 2008 WIT Press
www.witpress.com, ISSN 1743-3509 (on-line)

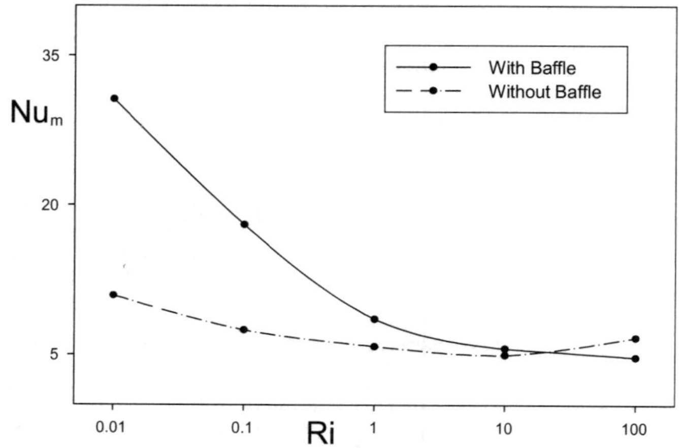

Figure 6: Average Nusselt number of the cavity.

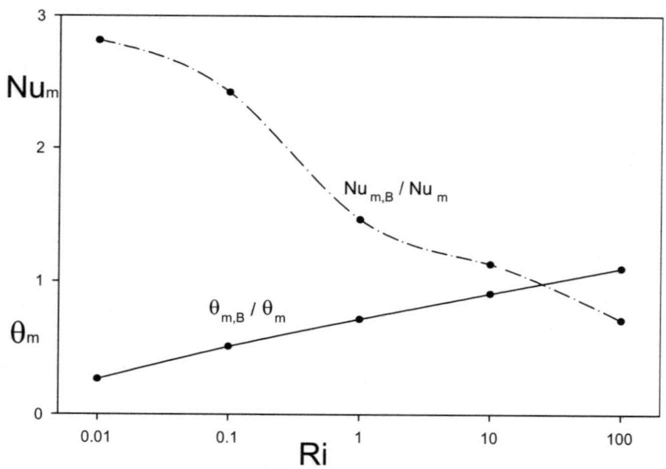

Figure 7: Average air temperatures and average Nusselt numbers ratios.

The results obtained at Ri=100 in terms of cavity average temperature and the average Nusselt number can be related to the fluctuating results of the flow field at this Richardson number. In fact, no steady solution was found due to the instability in the physical system for this situation. Other researchers also found the fluctuating results in closed and open cavities and discussed that the Hopf bifurcations lead to oscillating flows [13, 14]. In this work, the examination of heat transfer at Ri=100 is based on solving the unsteady equations using dimensionless time step, $\Delta\tau = 0.1$. Fig. 8 shows the time variation of the average cavity air temperature, θ_m for the two channels. It can be seen that the channel with baffle has a lower frequency and higher amplitude fluctuations. The steady state values of θ_m and Nu_m, presented in Figs 5, 6, and 7, are the mean time value in the asymptotic periodic zone.

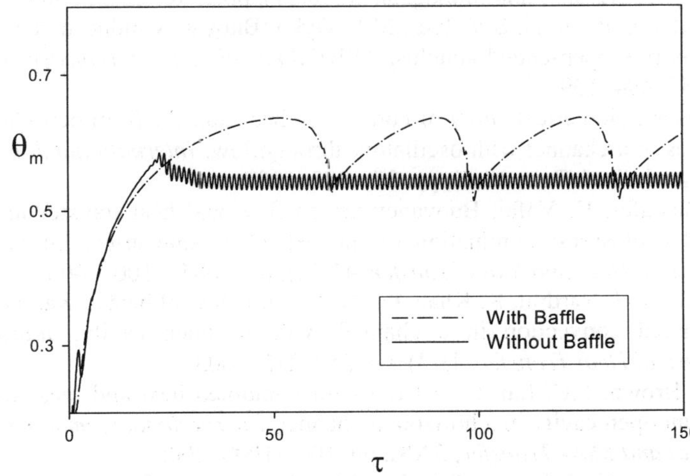

Figure 8: Time variation of average air temperature in cavity at Ri=100.

6 Conclusions

In this study, the results of a numerical modelling of mixed convection in an open cavity beneath of a two-dimensional channel with and without baffle are presented. The results of the average temperatures and average heat transfer have been performed for a series of Richardson numbers. The results show that, for low and medium Richardson numbers, the baffle has a key roll in the ventilation of the cavity; however, at a high Richardson number of Ri=100, the baffle has a negative effect on the cooling performance of the cavity.

References

[1] M. Hasnaoui, E. Bilgen, P. Vasseur, Natural convection above an array of open cavities heated from below, *Numerical Heat Transfer*, Part A, 18 pp. 463–482, 1990.

[2] G.F. Jones, J. Cai, Analysis of a transient asymmetrically heated/cooled open thermosyphon, *Journal of Heat Transfer*, 115 pp. 621–630,1993.

[3] R.A. Showole, J.D. Tarasuk, Experimental and numerical studies of natural convection with flow separation in upward-facing inclined open cavities, *Journal of Heat Transfer*, 115(3), pp. 592–605, 1993.

[4] S.W. Chang, S. Chiou, L. Su, T. Yang, Free connective heat transfer in tilted longitudinal open cavity, *Heat Transfer Engineering*, 26(10) pp.46-64, 2005.

[5] M. El Alami, M. Najam, E. Semma, A. Oubarra, F. Penot, Electronic components cooling by natural convection in horizontal channel with slots, *Energy Conversion and Management*, 46(17), pp. 2762-2772, 2005.

[6] K. Vafai, C.P. Desai, S.V. Iyer, M.P. Dyko, Buoyancy induced convection in a narrow open-ended annulus, *ASME Journal of Heat Transfer*, 119(3), pp. 483–494, 1997.

[7] T. Fusegi, Numerical study of convective heat transfer from periodic open cavities in a channel with oscillatory throughflow, *International Journal of Heat and Fluid Flow*, 18(4), pp. 376-383, 1997.

[8] K. Khanafer, K. Vafai, Buoyancy-driven flow and heat transfer in open-ended enclosures: elimination of the extended boundaries, International *Journal of Heat and Mass Transfer*, 43(22), pp. 4087–4100, 2000.

[9] O. Manca, S. Nardini, K. Khanafer, K. Vafai, Effect of heated wall position on mixed convection in a channel with an open cavity, *Journal of Numerical Heat Transfer*, 43(3), pp. 259–282, 2003.

[10] N.M. Brown, F.C. Lai, Correlations for combined heat and mass transfer from an open cavity in a horizontal channel, *International Communications in Heat and Mass Transfer*, 32(8), pp. 1000-1008, 2005.

[11] J.C. Leong, N.M. Brown, F.C. Lai, Mixed convection from an open cavity in a horizontal channel, *International Communications in Heat and Mass Transfer*, 32(5), pp. 583-592, 2005.

[12] O. Manca, S. Nardini, K. Vafai, Experimental investigation of mixed convection in a channel with an open cavity, *Experimental Heat Transfer*, 19(1), pp. 53-68, 2006.

[13] E. Papanicolaou, Y. Jaluria, Transition to a periodic regime in mixed convection in a square cavity, *Journal of Fluid Mechanics*, 239 pp. 489-509,1992.

[14] M.C. D'Orazio, C. Cianfrini, M. Corcione, Rayleigh–Bénard convection in tall rectangular enclosures, *International Journal of Thermal Sciences*, 43(2), pp. 135-144, 2004.

Parameter identification of the elastic modulus of ground rock based on blasting using the first order adjoint method

T. Ishimoto & M. Kawahara
Department of Civil Engineering, Chuo University, Japan

Abstract

In the case of tunnel excavation, it is important to investigate ground properties before staring construction. However, it takes much money and time to investigate the ground properties. Therefore, a numerical technique to solve the problem is proposed in this paper. The method is parameter identification, which is inverse analysis based on an optimal control theory. The main purpose of this research is to present the parameter identification of elastic modulus at the Kasakura tunnel site located in Fukui prefecture in Japan. The blasting is made at the tunnel face and the vibration of the blasting is measured at the observation points. These observation velocities are used for the performance function.
Keywords: finite element method, parameter identification, first order adjoint method, weighted gradient method.

1 Finite element equation

Applying the finite element method to the basic equation, the finite element equation discretized by the linear tetrahedral element is obtained as follows.

In this paper, indecial notation and summation convention with repeated indies are used. The basic equation is expressed by the following three equations; the equilibrium of stress equation, the strain-displacement equation and the stress-strain equation.

$$\sigma_{ij,j} - \rho b_i + \rho \ddot{u}_i = 0, \tag{1}$$

$$\varepsilon_{ij} = \frac{1}{2}(u_{i,j} + u_{j,i}), \tag{2}$$

WIT Transactions on the Built Environment, Vol 102, © 2008 WIT Press
www.witpress.com, ISSN 1743-3509 (on-line)
doi:10.2495/US080201

$$\sigma_{ij} = D_{ijkl}\varepsilon_{kl}, \tag{3}$$

where σ_{ij}, b_i, u_i, $\ddot{u}_i$ and ε_{ij} are total stress, body force, displacement, acceleration, and strain, respectively. The elastic stress – strain relation is expressed by D_{ijkl} and can be written as follows;

$$D_{ijkl} = \lambda\delta_{ij}\delta_{kl} + \mu(\delta_{ik}\delta_{jl} + \delta_{il}\delta_{jk}), \tag{4}$$

where δ_{ij} is Kronecker's delta, in which Lame's constants λ and μ are written as:

$$\lambda = \frac{\nu E}{(1-2\nu)(1+\nu)}, \tag{5}$$

$$\mu = \frac{E}{2(1+\nu)}, \tag{6}$$

where E and ν are the elastic modulus and the Poisson ratio, respectively. The basic equations are solved on the following boundary conditions.

$$u_i = \hat{u}_i^0 \qquad on \;\; t = t_0, \tag{7}$$

$$\dot{u}_i = \hat{u}_i^0 \qquad on \;\; t = t_0, \tag{8}$$

where $\hat{u}_i^0$ and $\dot{u}_i^0$ are the specified values given at the initial stage. The boundary Γ is divided into Γ_1 and Γ_2. On Γ_1 boundary, displacement is specified and on Γ_2 boundary, surface force t_i is given:

$$u_i = \hat{u}_i \qquad on \;\; \Gamma_1, \tag{9}$$

$$t_i = \sigma_{ij}n_j = \hat{t}_i \qquad on \;\; \Gamma_2, \tag{10}$$

where $\hat{u}_i$ and $\hat{t}_i$ are specified values on the boundary, n_j is the external unit normal to the boundary.

1.1 Finite element equation

Applying the finite element method to the basic equation, the finite element equation discretized by the linear tetrahedral element is obtained as follows;

$$M_{\alpha i\beta k}\ddot{u}_{\beta k} + C_{\alpha i\beta k}\dot{u}_{\beta k} + K_{\alpha i\beta k}u_{\beta k} = \hat{\Gamma}_{\alpha i}, \tag{11}$$

Each matrix can be written as follows;

$$M_{\alpha i\beta k} = \rho\int_V \delta_{ik}N_\alpha N_\beta dV, \tag{12}$$

$$K_{\alpha i\beta k} = \int_V N_{\alpha,j}D_{ijkl}N_{\beta,l}dV, \tag{13}$$

$$\Gamma_{\alpha i} = \int_V (N_\alpha \rho b_i) dV - \int_{\Gamma_2} (N_\alpha \hat{t}_i) d\Gamma, \tag{14}$$

The effect of damping can be expressed as follows;

$$C_{\alpha i\beta k} = \alpha_0 M_{\alpha i\beta k} + \alpha_1 K_{\alpha i\beta k}, \tag{15}$$

where N_α is the linear interpolation function of the finite element method. $M_{\alpha i\beta k}$ and $K_{\alpha i\beta k}$ are mass and elastic matrix, respectively.

2 Newmark β method

In this paper, the Newmark β method is applied to the discretization in time. In the Newmark β method, displacement and velocity at $(n+1)$ time are expressed as follows; $(n+1)$ time are expressed as follows;

$$u_{\beta k}^{(n+1)} = u_{\beta k}^{(n)} + \dot{u}_{\beta k}^{(n)} \Delta t + \frac{1}{2} \ddot{u}_{\beta k}^{(n)} \Delta t^2 + \beta \Delta t^2 (\ddot{u}_{\beta k}^{(n+1)} - \ddot{u}_{\beta k}^{(n)}), \tag{16}$$

$$\dot{u}_{\beta k}^{(n+1)} = \dot{u}_{\beta k}^{(n)} + \ddot{u}_{\beta k}^{(n)} \Delta t + \gamma \Delta t (\ddot{u}_{\beta k}^{(n+1)} - \ddot{u}_{\beta k}^{(n)}), \tag{17}$$

where β and γ are set as 0.25 and 0.50, respectively. Eqs. (16) and (17) are substituted into eq.(14), the following equations can be derived as follows;

$$G_{\alpha i\beta k} \ddot{u}_{\beta k}^{(n+1)} = \hat{\Gamma}_{\alpha i} - H_{\alpha i\beta k} \ddot{u}_{\beta k}^{(n)} - L_{\alpha i\beta k} \dot{u}_{\beta k}^{(n)} - K_{\alpha i\beta k} u_{\beta k}^{(n)}, \tag{18}$$

where $G_{\alpha i\beta k}$, $H_{\alpha i\beta k}$ and $L_{\alpha i\beta k}$ are written as:

$$G_{\alpha i\beta k} = M_{\alpha i\beta k} + \gamma \Delta t C_{\alpha i\beta k} + \beta \Delta t^2 K_{\alpha i\beta k}, \tag{19}$$

$$H_{\alpha i\beta k} = (1-\gamma) \Delta t C_{\alpha i\beta k} + (\frac{1}{2} - \beta) \Delta t^2 K_{\alpha i\beta k}, \tag{20}$$

$$L_{\alpha i\beta k} = C_{\alpha i\beta k} + \Delta t K_{\alpha i\beta k}, \tag{21}$$

Calculating acceleration $\ddot{u}_{\beta k}^{(n+1)}$ by eq.(18) and substituting it into eqs.(16) and (17), displacement $u_{\beta k}^{(n+1)}$ and velocity $\dot{u}_{\beta k}^{(n+1)}$ can be obtained.

3 Performance function

In this paper, the parameter identification is defined as finding optimal value so as to minimize the performance function. The performance function is defined as follows;

$$J = \frac{1}{2} \int_t (\dot{u}_{\alpha i} - \dot{u}_{\alpha i}^*) Q_{\alpha i\beta k} (\dot{u}_{\beta k} - \dot{u}_{\beta k}^*) dt, \tag{22}$$

where $\dot{u}_{\alpha i}$ and $\dot{u}_{\alpha i}^*$ are the computed and observed velocities and $Q_{\alpha i\beta k}$ is the weighting diagonal matrix.

4 First order adjoint equation

In this paper, the extended performance function J^* is expressed as follows;

$$J^* = \frac{1}{2}\int_t (\dot{u}_{\alpha i} - \dot{u}^*_{\alpha i}) Q_{\alpha i\beta k}(\dot{u}_{\beta k} - \dot{u}^*_{\beta k})dt \\ + \int_t \lambda^T{}_{\alpha i}(F_{\alpha i} - M_{\alpha i\beta k}\ddot{u}_{\beta k} - C_{\alpha i\beta k}\dot{u}_{\beta k} - K_{\alpha i\beta k}u_{\beta k})dt, \qquad (23)$$

where $\lambda_{\alpha i}$ is the Lagrange multiplier. Taking the first variation of the extended performance function, the gradient of the performance function, the adjoint equation and terminal condition can be calculated. The first variation of the extended performance function δJ^* is expressed as follows; The first variation of the extended performance function δJ^* is derived as;

$$\delta J^* = \int_t (\dot{u}_{\alpha i} - \dot{u}^*_{\alpha i}) Q_{\alpha i\beta k}\delta\dot{u}_{\beta k}dt \\ + \int_t \delta\lambda^T{}_{\alpha i}(F_{\alpha i} - M_{\alpha i\beta k}\ddot{u}_{\beta k} - C_{\alpha i\beta k}\dot{u}_{\beta k} - K_{\alpha i\beta k}u_{\beta k})dt \\ + \int_t \lambda^T(\delta F_{\alpha i} - M_{\alpha i\beta k}\ddot{u}_{\beta k} - C_{\alpha i\beta k}\dot{u}_{\beta k} - K_{\alpha i\beta k}u_{\beta k})dt, \qquad (24)$$

where δJ^* is transformed to obtain the gradient of performance function with respect the elastic modulus.

$$\delta J^* = \int_t (\dot{u}_i - \dot{u}^*_i) Q_{\alpha i\beta k}\delta\dot{u}_{\beta k}dt \\ + \int_t \delta\lambda^T{}_{\alpha i}(F_{\alpha i} - M_{\alpha i\beta k}\ddot{u}_{\beta k} - C_{\alpha i\beta k}\dot{u}_{\beta k} - K_{\alpha i\beta k}u_{\beta k})dt \\ + \int_t \lambda^T(\delta F_{\alpha i} - M_{\alpha i\beta k}\delta\ddot{u}_{\beta k} - C_{\alpha i\beta k}\delta\dot{u}_{\beta k} - K_{\alpha i\beta k}\delta\dot{u}_{\beta k} \\ - (\alpha_1 K^*{}_{\alpha i\beta k}u_{\beta k} + K^*{}_{\alpha i\beta k}\dot{u}_{\beta k})\delta E dt. \qquad (25)$$

where $K^* = K_{\alpha i\beta k}/E$. Integrating by parts, δJ^* is transformed as follows;

$$\delta J^* = (\dot{u}_{\alpha i}(t_f) - \dot{u}^*_{\alpha i}(t_f))Q_{\alpha i\beta k} - (\dot{u}_{\alpha i}(t_0) - \dot{u}^*_{\alpha i}(t_0))Q_{\alpha i\beta k} \\ - \lambda^T_{\alpha i}(tf)M_{\alpha i\beta k}\delta\dot{u}_{\beta k}(t_f) + \lambda^T_{\alpha i}(t_0)M_{\alpha i\beta k}\delta\dot{u}_{\beta k}(t_0) \\ + \dot{\lambda}^T_{\alpha i}(t_f)M_{\alpha i\beta k}\delta u_{\beta k}(t_f) + \dot{\lambda}^T_{\alpha i}(t_0)M_{\alpha i\beta k}\delta u_{\beta k}(t_0) \\ + \lambda^T_{\alpha i}(t_f)C_{\alpha i\beta k}\delta u_{\beta k}(t_f) + \lambda^T_{\alpha i}(t_0)C_{\alpha i\beta k}\delta u_{\beta k}(t_0) \\ - \int_t \ddot{\lambda}^T_{\alpha i}M_{\alpha i\beta k}\delta u_{\beta k}dt + \int_t \dot{\lambda}^T_{\alpha i}C_{\alpha i\beta k}\delta u_{\beta k}dt - \int_t \lambda^T_{\alpha i}K_{\alpha i\beta k}\delta u_{\beta k}dt \\ + \int_t \delta\lambda^T{}_{\alpha i}(F_{\alpha i} - M_{\alpha i\beta k}\ddot{u}_{\beta k} - C_{\alpha i\beta k}\dot{u}_{\beta k} - K_{\alpha i\beta k}u_{\beta k})dt$$

$$- \int_t (\ddot{u}_{\alpha i} - \ddot{u}^*_{\alpha i})^T Q_{\alpha i \beta j} \delta u_{\beta j} dt$$

$$- \int_t \lambda^T_{\alpha i} (\alpha_1 K^*{}_{\alpha i \beta k} u_{\beta k} + K^*{}_{\alpha i \beta k} \dot{u}_{\beta k}) \delta E dt, \tag{26}$$

The first variation of the extended performance function δJ^* should equal to 0. Then, the adjoint equation and the terminal condition are obtained.

$$M_{\alpha i \beta k} \ddot{\lambda}_{\beta k} - C_{\alpha i \beta k} \dot{\lambda}_{\beta k} + K_{\alpha i \beta k} \lambda_{\beta k} + (\ddot{u}_{\alpha i} - \ddot{u}^*_{\alpha i}) Q_{\alpha i \beta k} = 0. \tag{27}$$

$$\lambda_{\alpha i}(t_f) = 0, \tag{28}$$

$$M_{\alpha i \beta k} \dot{\lambda}_{\beta k}(t_f) + (\dot{u}_{\alpha i}(t_f) - \dot{u}^*_{\alpha i}(t_f)) Q_{\alpha i \beta j} = 0, \tag{29}$$

where $\ddot{\lambda}_{\beta k}(t_f)$ is the terminal condition of acceleration at the terminal time tf and $\ddot{\lambda}_{\beta k}(t_f)$ is solved for using $\dot{\lambda}_{\beta k}(t_f)$ and $\lambda_{\beta k}(t_f)$.

$$\begin{aligned} \ddot{\lambda}_{\beta k}(t_f) M_{\alpha i \beta k} &= -Q_{\alpha i \beta k} (\dot{u}_{\beta k} - \dot{u}^*_{\beta k}) \\ &\quad + \dot{\lambda}_{\beta k}(tf) C_{\alpha i \beta k} \\ &\quad - \lambda_{\beta k}(tf) K_{\alpha i \beta k}, \end{aligned} \tag{30}$$

The gradient of the performance function is derived as follows;

$$\mathrm{grad}(J^*)_{\beta k} = \lambda^T_{\alpha i} (\alpha_1 K^*{}_{\alpha i \beta k} \dot{u}_{\beta k} + K^*_{\alpha i \beta k} u_{\beta k}). \tag{31}$$

5 Weighted gradient method

The weighted gradient method is applied as the minimization technique. The modified performance function is expressed as follows;

$$K = J^* + \frac{1}{2} \int_t (X^{(n+1)}_\alpha - X^{(n)}_\alpha) W_{\alpha\beta} (X^{(n+1)}_\beta - X^{(n)}_\beta) dt, \tag{32}$$

where X_β is to be identified. The optimal condition of the modified performance function is as follows;

$$\delta K = 0. \tag{33}$$

The parameters can be updated at each iteration.

$$W_{\alpha\beta} X^{n+1}_\beta = W_{\alpha\beta} X^n_\beta - grad(J^*)_\beta. \tag{34}$$

Using eq.(34), the parameter is updated by the iterative calculations.

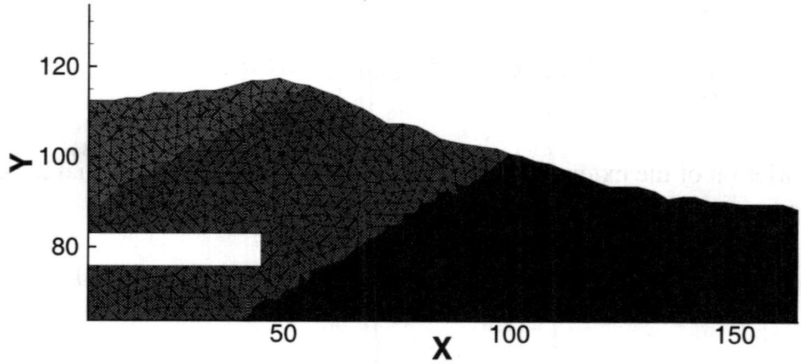

Figure 1: Finite element method.

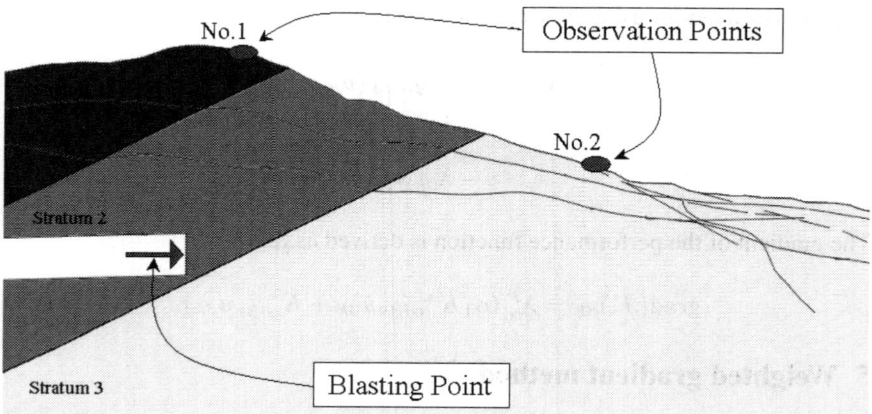

Figure 2: Computational domain.

6 Numerical study

The present method is applied to the Kasakura tunnel site. In the study, the actual observed data is employed. The purpose is to find parameter so as to minimize performance function.

The Kasakura tunnel construction site is located in Fukui prefecture in Japan. Fig.1 illustrates the finite element mesh. Total number of nodes and elements are 1045 and 1934, respectively. There are three layers in this area. The elastic modulus is considered to be unknown in layer 2. Blasting force and observed points are shown in Fig.2. This blasting force is assumed as $1.0 \times 10^8 [kN/m^2]$. The Poisson's ratio is 0.30. The observed points is set at No.1 and No.2. The elastic modulus in the stratum colored in layer 2 is identified.

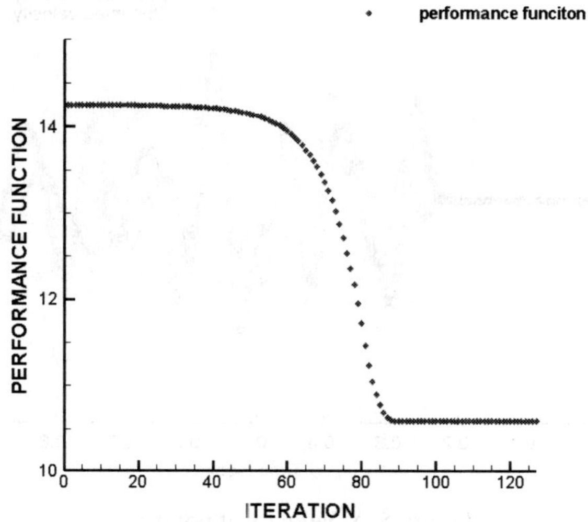

Figure 3: Performance function.

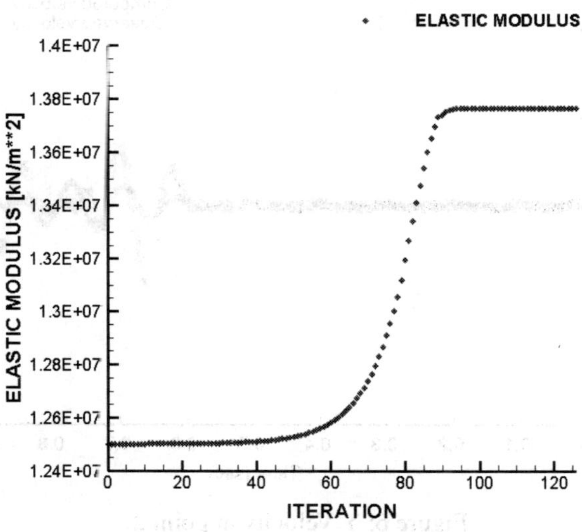

Figure 4: Elastic modulus.

6.1 Basic equation

The identification of the elastic modulus in the stratum colored in layer 2 is identified based on the velocities observed at observed points Nos. 1 and 2. Fig.3 shows the variation of performance function. Fig.4 shows the variation of the elastic mod-

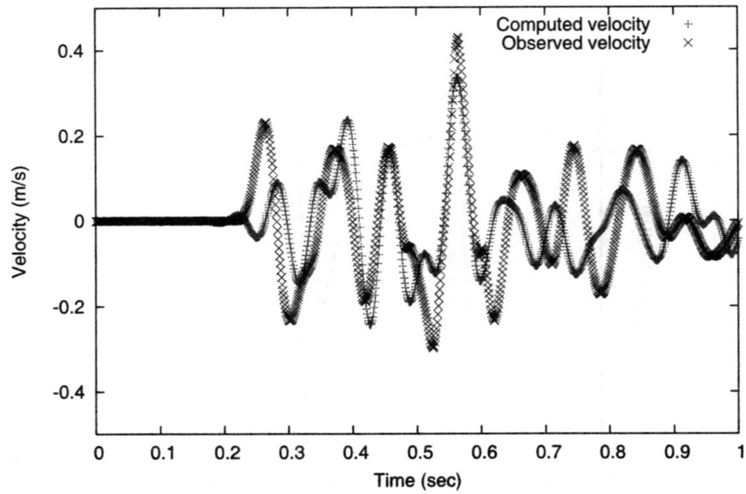

Figure 5: Y Velocity at point 1.

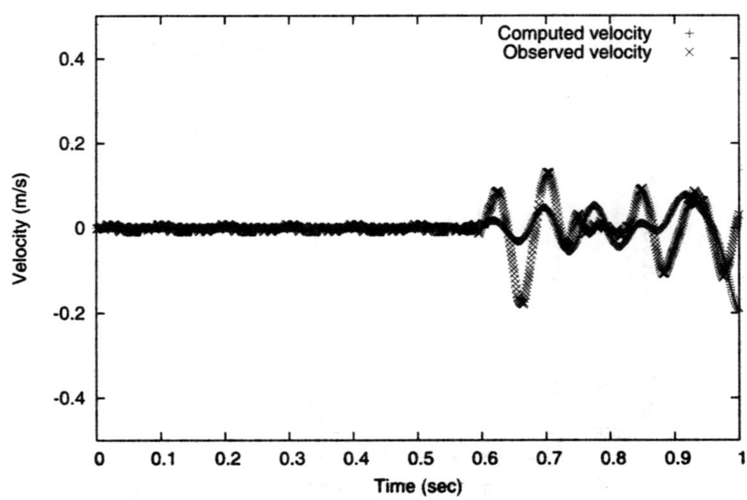

Figure 6: Y Velocity at point 2.

ulus. The final value of the elastic modulus computed based on the observed value is $1.38 \times 10^8 [kN/m^2]$, which is in close agreement with the value obtained by the drilling tests. Figs.5 and 6 show the variation of comparison velocity. Two data show almost same wave form. Few differences are influence of observed error and viscosity.

7 Conclusion

The elastic modulus can be identified using the finite element method and the first order adjoint method of an optimal control theory. The performance function is defined as the square sum of the difference between computed and observed velocities. The elastic modulus is found so as to minimize the performance function. The elastic modulus is converged to the target value staring from the initial value. The performance function is converged to 0. Thus, it is verified that the numerical method presented in this paper is correct and valid.

References

[1] I.M.Navon: Second-Order Information in Data Assimilation. *American Meteorological Society*, 629–648, (2002)

[2] R.H.F.Jackson and G.P.McCormick: Second-Order Sensitivity analysis in factorable programming.*Theory and applications. Math. Program*, 41, 1–28, (1988)

[3] R.H.F.Jackson and G.P.McCormick: The polyadic structure of factorable function tensors with application to high-order minimization techniques. *Journal of Optimization Theory and Applications*, 51, 63–94, (1986)

Author Index